BABURA

PARAMETER IDENTIFICATION FOR INDUCTION MOTOR DRIVES

BABURAJ KARANAYIL

PARAMETER IDENTIFICATION FOR INDUCTION MOTOR DRIVES

USING ARTIFICIAL NEURAL NETWORKS AND FUZZY LOGIC

VDM Verlag Dr. Müller

Impressum/Imprint (nur für Deutschland/ only for Germany)

Bibliografische Information der Deutschen Nationalbibliothek: Die Deutsche Nationalbibliothek verzeichnet diese Publikation in der Deutschen Nationalbibliografie; detaillierte bibliografische Daten sind im Internet über http://dnb.d-nb.de abrufbar.

Alle in diesem Buch genannten Marken und Produktnamen unterliegen warenzeichen-, marken- oder patentrechtlichem Schutz bzw. sind Warenzeichen oder eingetragene Warenzeichen der jeweiligen Inhaber. Die Wiedergabe von Marken, Produktnamen, Gebrauchsnamen, Handelsnamen, Warenbezeichnungen u.s.w. in diesem Werk berechtigt auch ohne besondere Kennzeichnung nicht zu der Annahme, dass solche Namen im Sinne der Warenzeichen- und Markenschutzgesetzgebung als frei zu betrachten wären und daher von jedermann benutzt werden dürften.

Coverbild: www.purestockx.com

Verlag: VDM Verlag Dr. Müller Aktiengesellschaft & Co. KG
Dudweiler Landstr. 99, 66123 Saarbrücken, Deutschland
Telefon +49 681 9100-698, Telefax +49 681 9100-988, Email: info@vdm-verlag.de
Zugl.: SYDNEY, The University of New South Wales, 2005

Herstellung in Deutschland:
Schaltungsdienst Lange o.H.G., Berlin
Books on Demand GmbH, Norderstedt
Reha GmbH, Saarbrücken
Amazon Distribution GmbH, Leipzig
ISBN: 978-3-639-20033-1

Imprint (only for USA, GB)

Bibliographic information published by the Deutsche Nationalbibliothek: The Deutsche Nationalbibliothek lists this publication in the Deutsche Nationalbibliografie; detailed bibliographic data are available in the Internet at http://dnb.d-nb.de .

Any brand names and product names mentioned in this book are subject to trademark, brand or patent protection and are trademarks or registered trademarks of their respective holders. The use of brand names, product names, common names, trade names, product descriptions etc. even without a particular marking in this works is in no way to be construed to mean that such names may be regarded as unrestricted in respect of trademark and brand protection legislation and could thus be used by anyone.

Cover image: www.purestockx.com

Publisher:
VDM Verlag Dr. Müller Aktiengesellschaft & Co. KG
Dudweiler Landstr. 99, 66123 Saarbrücken, Germany
Phone +49 681 9100-698, Fax +49 681 9100-988, Email: info@vdm-publishing.com

Printed in the U.S.A.
Printed in the U.K. by (see last page)
ISBN: 978-3-639-20033-1

ACKNOWLEDGEMENTS

I would like to acknowledge the relentless support from Professor Faz Rahman and Professor Colin Grantham for the research that I carried out at The University of New South Wales, which contributed to the manuscript of this book. I would like to thank those researchers whose valuable publications are referenced in this book.

Baburaj Karanayil

Dedicated to the memory of my parents

CONTENTS

ESTIMATOR AND FUZZY LOGIC

**COMBINED ANN ON-LINE ROTOR AND STATOR
RESISTANCE ESTIMATION**

LIST OF SYMBOLS

v_{ds}^s	d- axis stator voltage in stator reference frame
v_{qs}^s	q- axis stator voltage in stator reference frame
v_{dr}^s	d- axis rotor voltage in stator reference frame
v_{qr}^s	q- axis rotor voltage in stator reference frame
v_{ds}	d- axis stator voltage in synchronously rotating reference frame
v_{qs}	q- axis stator voltage in synchronously rotating reference frame
v_{dr}	d- axis rotor voltage in synchronously rotating reference frame
v_{qr}	q- axis rotor voltage in synchronously rotating reference frame
$\vec{v}_s^s, \vec{v}_s$	stator voltage vector in stator reference frame
i_{ds}^s	d –axis stator current in stator reference frame
i_{qs}^s	q –axis stator current in stator reference frame
i_{dr}^s	d –axis rotor current in stator reference frame
i_{qr}^s	q –axis rotor current in stator reference frame
i_{ds}	d –axis stator current in synchronously rotaing reference frame
i_{qs}	q –axis stator current in synchronously rotaing reference frame
i_{dr}	d –axis rotor current in synchronously rotaing reference frame
i_{qr}	q –axis rotor current in synchronously rotaing reference frame
$i_{ds}^{s\,*}$	d-axis stator current estimate in stator reference frame
$i_{qs}^{s\,*}$	q-axis stator current estimate in stator reference frame
$\vec{i}_s^s, \vec{i}_s$	measured stator current vector in stator reference frame
$\vec{i}_s^{s\,*}$	estimated stator current vector in stator reference frame
I_s	magnitude of the measured stator current vector
I_s^*	magnitude of the estimated stator current vector
i_L	load current

$\vec{\lambda}_r$ rotor flux linkage vector

$\vec{\lambda}_s$ stator flux linkage vector

λ_{rd} amplitude of rotor flux in synchrounously rotating reference frame

λ_{rdref} rotor flux reference in synchrounously rotating reference frame

λ_{ds}^s d-axis stator flux linkages in stator reference frame

λ_{qs}^s q-axis stator flux linkages in stator reference frame

λ_{ds} d-axis stator flux linkages in synchronously rotating reference frame

λ_{qs} q-axis stator flux linkages in synchronously rotating reference frame

λ_{dr}^s d-axis rotor flux linkages in stator reference frame

λ_{qr}^s q-axis rotor flux linkages in stator reference frame

λ_{dr} d-axis rotor flux linkages in synchronously rotating reference frame

λ_{qr} q-axis rotor flux linkages in synchronously rotating reference frame

$\lambda_{dr}^{s\,vm}$ d-axis rotor flux linkages estimated by voltage model in stator reference frame

$\lambda_{qr}^{s\,vm}$ q-axis rotor flux linkages estimated by voltage model in stator reference frame

$\lambda_{dr}^{s\,im}$ d-axis rotor flux linkage estimated by current model in stator reference frame

$\lambda_{qr}^{s\,im}$ q-axis rotor flux linkage estimated by current model in stator reference frame

$\lambda_{dr}^{s\,nm}$ d-axis rotor flux linkage estimated by neural network in stator reference frame

$\lambda_{qr}^{s\,nm}$ q-axis rotor flux linkage estimated by neural network in stator reference frame

R_s	stator resistance (Ω)
R_r	rotor resistance (Ω)
R'_s	stator resistance used in the controllers (Ω)
R'_r	rotor resistance used in the controllers (Ω)
$\hat{R}_s$	estimated stator resistance (Ω)
$\hat{R}_r$	estimated rotor resistance (Ω)
R_{r-est}	estimated rotor resistance (Ω)
ΔR_{r-est}	incremental change in estimated rotor resistance (Ω)
L_{ls}	stator leakage inductance, H
L_{lr}	rotor leakage inductance, H
L_m	magnetizing inductance, H
L_r	rotor inductance, H
L_s	stator inductance, H
T_r	rotor time constant
T_s	sampling period of controllers
σ	leakage coefficient $=\left(1-L_m^2/L_sL_r\right)$
P	number of poles
τ	time constant
G	gain
K_p	proportional gain
K_i	integral gain
W_1,W_2,W_3	neural network weights in rotor resistance estimator
W_4,W_5,W_6,W_7	coefficients in stator current estimation
E_1,E_2	cumulative error function
η_1,η_2	training coefficients
α_1,α_2	momentum constants
$\vec{\varepsilon}$	error function vector

$\varepsilon_1, \varepsilon_2$	error functions
J_T	moment of inertia, kgm^2
T_e	developed torque in Nm
T_L	load torque in Nm
ω_m	mechanical rotor speed in rev/minute
ω_r	electrical rotor speed in radians/second
ω, ω_e	stator frequency in radians/second
ω_r^{est}	estimated electrical rotor speed in radians/second
ω_{mref}	mechanical speed reference in rev/minute
ω_{sl}	slip speed in radians/second
ω_{est}	estimated stator frequency in radians/second
θ	flux angle in degree

Superscripts

*	reference or command value
$\wedge$	estimated value

Subscrpits

ref	reference value
0	initial value

Abbreviation

IM	Induction motor
IFOC	Indirect Field Oriented Control
VSI	Voltage Source Inverter
RFOC	Rotor Flux Oriented Control
ANN	Artificial Neural Network
PI	Proportional Integral
LPF	Low-Pass Filter
SRE	Stator Resistance Estimator
RRE	Rotor Resistance Estimator

PWM	Pulse Width Modulation
SCIM	Squirrel-Cage Induction Motor
SRIM	Slip-Ring Induction Motor
VMFO	Voltage Model Flux Observer
emf	electromotive force

CHAPTER 1

INTRODUCTION

1.1 Parameter identification and control of rotor flux oriented induction motor drives

During the past three decades, the adjustable speed AC drive technology has gained a lot of momentum. It is well recognized that AC motor drives accounted for more than 50% of all electrical energy consumed in developed countries. Induction motors, wound rotor synchronous motors and permanent magnet synchronous motors are the best candidates for variable speed drive applications. The wound field synchronous motors have brushes, which result in wear and tear and needs regular maintenance. The permanent magnet synchronous motors are much more expensive even today because of the higher cost of permanent magnets which make them uneconomical in many applications. Even though squirrel-cage induction motors are the cheapest, easy to maintain and manufactured up to several mega watts power ratings, the modeling and control of such motors are complex compared to the other motors.

Electromechanical drives are used for the overwhelming majority of industrial actuation applications. Most such applications require the control of the position, speed or torque of the electric motor or the mechanical actuator. The combination of an electric motor, a power electronic controller, a mechanical transmission device and a feedback control system is commonly known as a servomechanism. Motion control can be defined as the application of high performance servo drives to the control of torque, speed and/or position. Even though the field of motion control stemmed from the area of power electronics initially, it has acquired a separate status of its own now. The early stages of research in this area, in the 1970s, focussed on the application of basic control theory to drive a motor, using a power electronic converter. As the field evolved, it drew in researchers who originally trained in various other fields, including control engineering, mechanical engineering, artificial

1

intelligence (including fuzzy logic and artificial neural networks) and electrical machines. There are two central issues and problems in motion control. One is to make the resulting system of controller and plant robust against parameter variations and disturbances. The other is to make the system intelligent, that is to make the system self adjusting to changes in environment and system parameters. Various methods in control systems theory have been applied to improve the robustness of a motor control system. To make a motion control system intelligent, requires soft computational methods such as artificial neural networks (ANNs) and fuzzy logic. These are familiar intelligent control methods, and they have been applied by many researchers working in the drives area. In the future, it is expected that computational power will lead to realization of more powerful control techniques. It will be possible to make a totally automatic control system which will obviate the need for a plant model by using an intelligent control technique like ANNs. It will be possible to identify the required parameters, decide on the control strategy and self-commission the drive.

Perhaps the greatest progress in recent years among electric motor control systems is in the development of induction motor drives. This can be attributed mainly to the invention of field oriented control by Blaschke in 1972 [1], the development of direct torque control by Depenbrock [2] and Takahashi [3] in 1985 and the rapid development of digital and power electronics in the last two decades.

Traditionally, there are two typical field orientation control approaches for induction motor drives: Direct Field Orientation (DFO) control and Indirect Field Orientation (IFO) control as discussed by Bose [4]. In induction motor drives using DFO, the motor flux is calculated using measured voltages and currents with some known parameters of the machine. This scheme leads to poor accuracy because of the machine parameter dependency and the use of pure integration for the flux calculation and hence is seldom employed in industrial drives.

IFO induction motor drives have been used for numerous industrial applications in the last decade. Instead of observing the machine flux, the correct field orientation control is obtained by a feedforward slip control in an IFO based drive. With the use of a shaft encoder, the Rotor Flux Oriented Control (RFOC) can be accomplished with low cost and high performance. However, the accuracy of the slip calculation used in RFOC also depends significantly on the rotor time constant T_r of the machine, which varies with changes in temperature and the load of the machine. Many parameter identification schemes have been studied to minimize the detuning effects and to enhance the drive system performance. These methods are reviewed in detail in Chapters 2 and 3.

The induction motor by itself is very robust for use in harsh environments. To obtain precise speed control, a speed sensor, typically an optical shaft encoder is normally used in induction motor drives. The presence of such a sensor brings several disadvantages from the standpoint of robustness, drive cost, reliability, machine noise and noise immunity. Elimination of the speed sensor together with parameter identification is thus an active area of both academic and industrial research.

1.2   Scope of this project

In Indirect Field Orientation of induction motors, the major problem is variation in the rotor resistance which is affected by change in rotor temperature. The practical temperature excursion of the rotor is approximately 130°C above ambient. This increases the rotor resistances by 50 percent over its ambient or nominal value. When this parameter is incorrect in the controller, the calculated slip frequency is incorrect and the flux angle is no longer appropriate for field orientation. This results in instantaneous error in both flux and torque which can be shown to excite a second order transient characterized by an oscillation frequency equal to the command slip frequency. The rotor flux could rise by 20% in theory, but practically this can not happen due to magnetic saturation. There can also be 20% error in the amplitude of steady-state torque since the steady-state slip is also incorrect. In addition, steady-state slip errors also cause additional motor heating and reduced efficiency.

The effect of change in rotor temperature in the IFO induction motor drive is revisited by experiment. The experimental results of speed, torque and rotor flux linkage of a rotor flux oriented induction motor drive operating at two different temperatures which significantly vary the rotor resistance are shown in Figures 1.1 and 1.2. Figure 1.1 is for the drive, when the motor is cold (at 25°C), when the drive controllers were initialized with the cold parameters of the motor. Figure 1.2 is for the same machine operating at 75°C for more than an hour, keeping the motor parameters used in the controller unaltered. It is clear that the drive has started to oscillate, because of the error between the real motor torque and the torque used by the controllers. This difference in torque is in turn caused by the error in the actual rotor resistance and the rotor resistance used in the feedforward controller in the RFOC. The detuning effect due to the mismatch of the rotor time constant T_r in the motor and RFOC is thus a major problem.

Methods of overcoming this problem by a number of ways have been a major research goal for a number of years. The main objective of this thesis is to analyse, develop and implement a very fast on-line parameter identification algorithm using artificial neural networks for use in rotor flux oriented vector control, to eliminate the problems such as shown in Figure 1.2.

Many researchers have brought out methods of identifying rotor time constant in RFOC for induction motor drives. There was one class of methods estimating the rotor time constant using the *spectral analysis techniques*. This group of methods is based on the measured response to a deliberately injected test signal or an existing characteristic harmonic in the voltage/current spectrum. Stator currents and / or voltages of the motor are sampled and the parameters are derived from the spectral analysis of these samples. The second classification of rotor resistance identification schemes used *observer based techniques*. Most of these methods have used the Extended Kalman filter, which are computational intensive techniques. The third group of on-line rotor resistance adaptation methods is based on principles of *model*

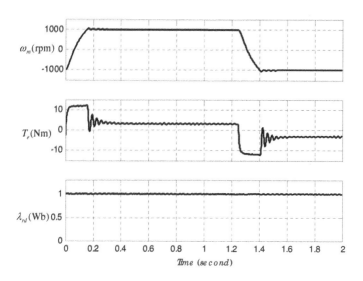

Figure 1.1 Experimental results- full load reversal with $R'_r = R_r$ and stator temperature: 25°C.

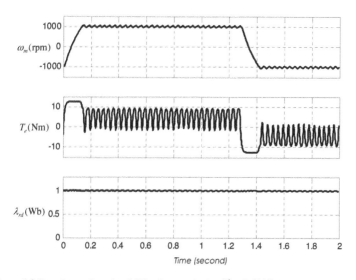

Figure 1.2 Experimental results- full load reversal with $R'_r = 0.611R_r$ and stator temperature: 75°C

reference adaptive control. This is the approach that has attracted most of the attention due to its relatively simple implementation requirements. One of the

5

common features that all of the methods of this group share is that rotor resistance adaptation is usually operational in steady-states only and is disabled during transients. In addition to the above methods, there are also a few techniques proposed which can not be classified in the above three categories. Even these identification methods are valid only for steady-state operation of the induction motor.

Another possibility, opened by the recent developments in artificial intelligence, is the application of artificial neural networks and fuzzy logic for the on-line rotor time constant / rotor resistance adaptation. Fuzzy logic is simple to implement and estimators using fuzzy logic have no convergence issues with the estimators using fuzzy logic. For these reasons fuzzy rotor resistance adaptation has been attempted in the past.

The artificial neural network estimator has the advantages of faster execution speed and fault tolerant characteristics compared to the estimators implemented in a DSP based system. They have the attributes of estimating parameters of a non-linear system with good accuracy and fast response. In this thesis, the above benefits of using an ANN for parameter identification have been utilized to adapt the rotor resistance in a rotor flux oriented induction motor drive.

1.3 Outline of the book

This book begins with a review of induction motor parameter estimation techniques in rotor flux oriented control in chapter 2. The principle of both Direct Field Orientation and Indirect Field Orientation and the influence of parameter errors are first discussed. The available literature covering on-line rotor time constant estimation techniques, broadly classified as spectral analysis techniques, observer based techniques, model reference adaptive system based techniques and the heuristic methods are reviewed in chapter 2. This chapter also reviews estimation of the stator resistance of the induction motor, because the stator flux used in the speed sensorless drive is detuned by the variation in the stator resistance. Also, some of the rotor resistance identification algorithms were dependent on the stator resistance of the motor. Some of these stator resistance identification schemes reported hitherto are reviewed in chapter 2.

In chapter 3, the background theory of conventional, fuzzy and neural function approximation schemes are included. The existing literature on applications of artificial neural networks for estimation of rotor flux, torque and rotor speed in induction motor drives is briefly reviewed.

Analysis and practical implementation of a rotor flux oriented induction motor drive is discussed in chapter 4. A rotor resistance identification algorithm has been verified with a simple PI estimator and a fuzzy estimator. Some modeling and experimental results are presented, to verify their capabilities.

An alternative method for rotor resistance identification using Artificial Neural Networks is discussed in chapter 5. The derivation of new mathematical models and performance analysis of this type of estimation using SIMULINK is presented. The validity of modeling results is verified by conducting experiments on a slip-ring and a squirrel-cage induction motor, and these are described.

The rotor resistance estimation using ANN is implemented with the help of rotor flux estimated using the voltage model of the induction motor. This rotor flux is computed from the measured stator currents and the stator flux estimated from the induction motor voltage model. This voltage model stator flux estimation is carried out using a programmable cascaded low-pass filter, which is investigated in detail in chapter 6. The experimental set up for implementation, analysis results and experimental results are discussed in detail.

Because of the use of a voltage model based estimation for rotor flux, the variation in stator resistance is also known to introduce errors in rotor resistance estimates. On-line stator resistance estimation thus has the potential to estimate R_r more accurately and to compensate for their variations. At first a simple PI compensator was investigated for the stator resistance estimation for the induction motor drive. The modeling and experimental results are presented for this estimator in chapter 7. Because the tuning of the PI controller proved to be difficult, an alternative compensator using a fuzzy logic system was proposed. This fuzzy estimator overcame the problem encountered with the PI estimator. The development of the fuzzy estimator, its analysis and experimental results are presented in chapter 7.

Yet another alternative stator resistance estimation technique using ANN was developed which is based on measured instantaneous currents and requires shorter real-time computation time. The mathematical modelling of this method and results of experimental investigation are presented in chapter 8.

The combined on-line rotor and stator resistance estimations discussed in chapter 8 was found to estimate rotor fluxes accurately with a speed sensor. The rotor speed could then be estimated using the induction motor state equations. These rotor and stator resistance estimators were also investigated in the case of a speed sensorless operation of the induction motor. The modeling and experimental results for this situation is presented in chapter 9. The estimated speed was found to follow very closely the actual speed obtained from the encoder.

Finally, conclusions and suggestions for future work are given in chapter 10.

CHAPTER 2

REVIEW OF INDUCTION MOTOR PARAMETER ESTIMATION TECHNIQUES WITH ROTOR FLUX ORIENTED CONTROL

2.1 Introduction

This chapter starts with a brief review of field oriented control techniques for the induction motor and then investigates the parameters which critically affect such controls. The variations of the rotor time constant L_r/R_r are investigated in terms of its extent and causes, and the importance of tracking this parameter and the methods of tracking developed up to date are reviewed. The control context in which stator resistance variation, in addition to rotor resistance variation, is needed is brought out and the methods of stator resistance tracking developed up to date are also reviewed.

2.2 Field oriented control

The first major breakthrough in the area of high performance induction motor drives came with the discovery of the concept of Field Oriented Control (FOC) by Blaschke [1] in 1972. Blaschke examined how field orientation occurs naturally in a separately excited dc motor. The armature and field fluxes are always perpendicular to each other in the dc motor. In an induction machine, a similar condition can be created with appropriate control of stator currents in the synchronously rotating frame of reference.

FOC is a technique which provides a method of decoupling the two components of stator current, one producing the flux and the other producing the torque. Therefore it provides independent control of torque and flux during both dynamic and steady state conditions. In FOC, the stator phase currents are transformed into a synchronously rotating reference frame and field orientation is achieved by aligning the rotor flux vector along the *d*-axis of the synchronously rotating reference frame. Figure 2.1 shows how the stator current vector can be decoupled along the rotor flux vector. The *d-q* axis model of the induction motor with the reference axes rotating at synchronous speed ω is given by Bose [4].

$$\vec{v}_s = R_s \vec{i}_s + \frac{d}{dt}\left(\vec{\lambda}_s\right) + j\omega\vec{\lambda}_s \tag{2.1}$$

$$0 = R_r \vec{i}_r + \frac{d}{dt}\left(\vec{\lambda}_r\right) + j\left(\omega - \omega_r\right)\vec{\lambda}_r \tag{2.2}$$

$$T_e = \frac{3}{2}p\frac{L_m}{L_r}\left(\lambda_{dr}i_{qs} - \lambda_{qr}i_{ds}\right) \tag{2.3}$$

$$J_T\frac{d\omega_m}{dt} = \frac{J_T}{p}\frac{d\omega_r}{dt} = T_e - T_L \tag{2.4}$$

where

$$\vec{v}_s = v_{ds} + jv_{qs} \tag{2.5}$$

$$\vec{i}_s = i_{ds} + ji_{qs} \tag{2.6}$$

$$\vec{i}_r = i_{dr} + ji_{qr} \tag{2.7}$$

$$\vec{\lambda}_s = \lambda_{ds} + j\lambda_{qs} \tag{2.8}$$

$$\vec{\lambda}_r = \lambda_{dr} + j\lambda_{qr} \tag{2.9}$$

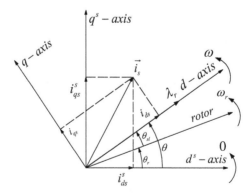

Figure 2.1 Transformation of stator current vector.

In the above equations, λ refers to the flux linkages and these quantities are expressed in the synchronously rotating reference frame, the d and q in the subscript stand for the direct and quadrature axes respectively and s and r in the subscript stand for the stator and rotor quantities respectively.

The field orientation concept implies that the current components supplied to the machine should be oriented in phase (flux component) and in quadrature (torque component) to the rotor flux vector $\overline{\lambda_r}$ and locking the phase of the reference system such that the rotor flux is entirely in the d-axis (flux axis), resulting in the mathematical constraint $\lambda_{qr} = 0$.

It should be noted that the control is performed on quantities obtained in the synchronous reference frame and when the rotor flux vector is chosen for decoupling, the control scheme is normally referred to as the Rotor Flux Oriented Control (RFOC).

With this arrangement, the control dynamics of the highly nonlinear structure of the induction motor becomes linearized and decoupled, Ho and Sen [5]. The two basic schemes of field orientation are:

- Direct Field Oriented control (DFO)
- Indirect Field Oriented control (IFO)

Direct Field Orientation originally proposed by Blaschke [1], requires flux acquisition which is mostly obtained from computational techniques using machine terminal quantities. Whereas IFO, proposed by Hasse [6], avoids the direct flux acquisition, by adding an estimated and regulated slip frequency to the shaft speed and integrating the total to obtain the stator flux position.

2.2.1 Direct field oriented control

In DFO the position of the stator flux, which is essential for the correct orientation, is directly measured using search coils or estimated from terminal measurements.

However using sensors to acquire the flux information makes it impossible to use off-the shelf induction motors because installation of such sensors can be done only during machine manufacturing. Instead, the measured terminal quantities such as stator voltages and currents can be used. The block diagram of such a system is shown in Figure 2.2. To estimate the rotor flux, one of the possible ways is to use the machine equations as an open loop observer using voltage equations in (2.10) – (2.13). Gabriel *et al* [7] estimated the rotor flux using these equations and these are commonly known as Voltage Model Flux Observer (VMFO). At very low frequencies (including zero speed) the control becomes difficult with the voltage model based estimation, as the voltages are very low and variations in stator resistance due to temperature rise and switch voltage drops, dead times etc tend to reduce the accuracy of the estimated signals. One of the solutions to this problem is to use current model equations for flux estimation at low speeds. There should be a proper mechanism to switch between the voltage and current models for successful operation. A variety of flux observers can be employed to obtain improved response and reduced sensitivity to machine parameters [8], [9], [10], [11] and [12].

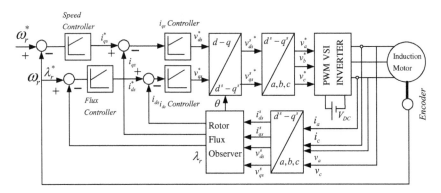

Figure 2.2 Direct Field Oriented drive system.

$$\lambda_{ds}^s = \int \left(v_{ds}^s - R_s i_{ds}^s \right) dt \qquad (2.10)$$

$$\lambda_{qs}^s = \int \left(v_{qs}^s - R_s i_{qs}^s \right) dt \qquad (2.11)$$

$$\lambda_{dr}^{s} = \frac{L_{r}}{L_{m}}\left(\lambda_{ds}^{s} - \sigma L_{s} i_{ds}^{s}\right) \tag{2.12}$$

$$\lambda_{qr}^{s} = \frac{L_{r}}{L_{m}}\left(\lambda_{qs}^{s} - \sigma L_{s} i_{qs}^{s}\right) \tag{2.13}$$

The scheme proposed by Jansen *et al* [8] reduces the dependence of rotor flux estimation on the stator resistance by using a current model observer at low speed, where the stator resistance effect is more significant. At high speed, the rotor flux is estimated using a voltage model flux observer where the stator resistance effect is reduced. There should be a proper mechanism to switch between the voltage and current models for successful operation. A variety of flux observers can be employed to obtain improved response and reduced sensitivity to machine parameters. Jansen *et al* [9] improved the observer performance by using closed loop rotor flux observers which used the estimated stator current error.

Also, flux observers have been designed by Yen *et al* [10] and Benchaib *et al* [11], using the sliding model technique for speed sensorless control of induction motors. Rehman *et al* [12] proposed a current model flux observer using a new sliding mode technique. In the current and flux observers all the terms that contain rotor time constant and rotor speed have been replaced by the sliding mode functions, thus the proposed current and flux estimations were completely insensitive to rotor time constant variation and any error in the estimated speed. In spite of their modification with the sliding mode controller, the drive could go down only up to ±5% of the rated speed.

The closed-loop velocity invariant, flux observer with current model input has the desirable low speed attributes of the current model, and the desirable high speed attributes of the voltage model. Hence these observers are ideally suitable for wide speed range applications requiring both zero speed and field weakening operation.

2.2.2 Indirect field oriented control

The indirect field oriented control method is essentially the same as direct field orientation, except that the flux position θ in Figure 2.1 is generated in the feedforward manner.

The rotor circuit equations of the induction motor can be written as:

$$\frac{d\lambda_{dr}}{dt} + R_r i_{dr} - (\omega - \omega_r)\lambda_{qr} = 0 \tag{2.14}$$

$$\frac{d\lambda_{qr}}{dt} + R_r i_{qr} + (\omega - \omega_r)\lambda_{dr} = 0 \tag{2.15}$$

The rotor flux linkage expressions can be written as:

$$\lambda_{dr} = L_r i_{dr} + L_m i_{ds} \tag{2.16}$$

$$\lambda_{qr} = L_r i_{qr} + L_m i_{qs} \tag{2.17}$$

The rotor currents can be written from (2.16) and (2.17) as:

$$i_{dr} = \frac{1}{L_r}\lambda_{dr} - \frac{L_m}{L_r}i_{ds} \tag{2.18}$$

$$i_{qr} = \frac{1}{L_r}\lambda_{qr} - \frac{L_m}{L_r}i_{qs} \tag{2.19}$$

Substituting the rotor current equations (2.18) and (2.19) into (2.14) and (2.15), they become:

$$\frac{d\lambda_{dr}}{dt} + \frac{R_r}{L_r}\lambda_{dr} - \frac{L_m}{L_r}R_r i_{ds} - \omega_{sl}\lambda_{qr} = 0 \tag{2.20}$$

$$\frac{d\lambda_{qr}}{dt} + \frac{R_r}{L_r}\lambda_{qr} - \frac{L_m}{L_r}R_r i_{qs} - \omega_{sl}\lambda_{dr} = 0 \tag{2.21}$$

Where, $\omega_{sl} = \omega - \omega_r$.

For decoupling control, it is desirable that

$$\lambda_{qr} = 0 \tag{2.22}$$

This implies,

$$\frac{d\lambda_{dr}}{dt} = 0 \tag{2.23}$$

so that the total rotor flux λ_r is directed along the *d*-axis.

Substituting the above conditions in Equations (2.20) and (2.21), we get

$$\frac{L_r}{R_r}\frac{d\lambda_r}{dt} + \lambda_r = L_m i_{ds} \tag{2.24}$$

$$\omega_{sl} = \frac{L_m}{\lambda_r}\frac{R_r}{L_r} i_{qs} \tag{2.25}$$

Where $\lambda_r = \lambda_{dr}$ has been substituted.

If rotor flux λ_r =constant, then from (2.24),

$$\lambda_r = L_m i_{ds} \tag{2.26}$$

Then the slip signal can be written as:

$$\omega_{sl} = \frac{1}{T_r}\frac{i_{qs}}{i_{ds}} \tag{2.27}$$

The torque developed by the induction motor is given by:

$$T_e = \frac{3}{2}\frac{P}{2}\left(\lambda_{dr} i_{qr} - \lambda_{qr} i_{dr}\right) \tag{2.28}$$

Because of decoupled control, from (2.22), (2.28) can be written as follows:

$$T_e = \frac{3}{2}\frac{P}{2}\lambda_{dr} i_{qr} \tag{2.29}$$

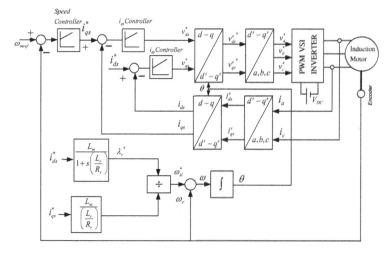

Figure 2.3 Indirect Field Oriented drive system.

Using (2.19) to replace i_{qr} in (2.29), it follows:

$$T_e = \frac{3}{2}\frac{P}{2}\lambda_{dr}\left(\frac{1}{L_r}\lambda_{qr} - \frac{L_m}{L_r}i_{qs}\right) \tag{2.30}$$

$$T_e = \frac{3}{2}\frac{P}{2}\frac{L_m}{L_r}\lambda_{dr}i_{qs} \tag{2.31}$$

An alternative to the direct sensing of flux position is to employ the slip relation, given in eqn (2.27), to estimate the flux position. Figure 2.3 illustrates this concept and shows how the rotor flux position can be obtained by integrating the sum of the rotor speed and the command slip frequency calculated using eqn (2.27). In the steady state this corresponds to setting the slip to the specific value which divides the stator current into flux producing and torque producing components. Indirect field orientation does not have inherent low speed problems and is thus preferred in most systems which must operate near zero speed.

2.2.3 Influence of parameter errors

In IFO, the major problem is the rotor resistance which is sensitive to temperature. The practical temperature excursion of the rotor is approximately 130°C above ambient [13]. This increases the rotor resistances by 50 percent over its ambient or nominal value. When this parameter is incorrect in the controller, the calculated slip frequency based on equation (2.27) is incorrect and the flux angle is no longer appropriate for field orientation. This results in instantaneous error in both flux and torque which can be shown to excite a second order transient characterized by an oscillation frequency equal to the command slip frequency. The rotor flux could rise by 20% in theory, but practically this can not happen due to magnetic saturation [13]. There can also be 20% error in the amplitude of steady-state torque since the steady-state slip is also incorrect [13]. In addition, steady-state slip errors also cause additional motor heating and reduced efficiency.

The detuning effect of RFOC due to the temperature rise of the motor has already been explained in section 1.2. The waveforms in Figure 1.2 were recorded when the drive was found to oscillate and the corresponding stator temperature was recorded as 75°C. Later on, when the rotor resistance identification described in section 4.4 was implemented with a PI estimator, it was found that the rotor resistance was increased by 63.6% at the end of the heat-run. This implies that the rotor resistance used in the controller R'_r was 0.611 R_r, when the drive was found to oscillate in section 1.2 for the Figure 1.2.

DFO systems are generally sensitive to stator resistance and total leakage but the various systems have individual detuning properties. Typically, parameter sensitivity is less than in IFO, especially when a flux regulator is employed. In all cases, both direct and indirect, parameter sensitivity depends on the *L/R* ratio of the machine with larger values giving greater sensitivity. Thus large high efficiency machines tend to have higher sensitivity to parameter errors.

Both basic types of field orientation have sensitivity to machine parameters and provide non-ideal torque control characteristics when control parameters differ from the actual machine parameters. In general, both steady-state and dynamic responses of torque differ from the ideal instantaneous torque control achieved in theory by a correctly tuned controller.

2.3   Status of rotor time constant estimation techniques

The on-line methods of rotor resistance identification methods developed so far could be broadly classified under the following categories:

- Spectral analysis techniques
- Observer based techniques
- Model reference adaptive system based techniques
- Heuistic methods

2.3.1 Spectral analysis techniques

This group of methods is based on the measured response to a deliberately injected test signal or an existing characteristic harmonic in the voltage/current spectrum. Stator currents and/or voltages of the motor are sampled and the parameters are derived from the spectral analysis of these samples. In the case of spectral analysis, a perturbation signal is used because under no load conditions of the induction motor, the rotor induced currents and voltages become zero, so slip frequency becomes zero, and hence, the rotor parameters cannot be estimated. In systems proposed by Matsuo and Lipo [14] and Toliyat and Hosseiny [15], the disturbance to the system is provided by injecting negative sequence components. Matsuo and Lipo [14] proposed an on-line technique for determining the value of rotor resistance by detecting the negative sequence voltage. The main drawback of this method is that the strong harmonic torque pulsation is induced due to the interaction of positive and negative rotating components of MMF.

Toliyat and Hosseiny [15] presented another on-line estimation technique based on the *d-q* model in the frequency domain. The *q*-axis component of the injected negative sequence component is kept at zero, so that the motor torque is undisturbed. The *d*-axis component affects the motor flux. FFT is used to analyse the currents and voltages and the fundamental components of the sampled spectral values are used to determine the parameters. Average speed is used for the identification of parameters. Gabriel and Leonard [16] proposed a correlation method to detect misalignment between the actual motor flux and the rotor flux given by the model. A small auxiliary signal is added to the *d*-axis flux component of the stator current and a correlation function is evaluated. The nonzero value of the correlation function indicates both coupling between fluxes and discrepancies between the parameters of the model and those the motor. In the method proposed by Sugimoto and Tamai [17], a sinusoidal perturbation is injected into the flux producing stator current component. Though rotor resistance can be estimated under any load and speed condition, the cost is high due to the installation of two search coils and will not be applicable to any off-the shelf induction motor.

2.3.2 Observer based techniques

The second classification of rotor resistance identification methods used observer based techniques. Lipo *et al* [18] proposed a method for detection of the inverse rotor time constant using the Extended Kalman Filter (EKF) by treating the rotor time constant as the fifth state variable along with the stator and rotor currents. Here, the wide band harmonics contained in the PWM inverter output voltage serve as the perturbation. This method works on the assumption that when the motor speed changes, the machine model becomes a two input/two output time varying system with superimposed noise input. The drawbacks are that this method assumes that all other parameters are known and the magnetizing inductance can introduce large errors into the rotor time constant estimation. Acarnley *et al* [19] proposed an application of EKF for slip calculation for tuning an IFO drive. Here, the Riccati difference equation is replaced by a look up table. Although the complexity of Riccati equation is reduced, the full-order EKF is computationally very intensive.

2.3.3 Model reference adaptive system based techniques

The third group of on-line rotor resistance adaptation methods is based on principles of model reference adaptive control. This is the approach that has attracted most of the attention due to its relatively simple implementation requirements. Here the basic idea is to estimate certain states from two different directions, one is to calculate using states of the controllers and the other is to estimate the same states using measured signals. One of the estimates should be independent of rotor resistance, so that the error between these two estimates provide the correction to the rotor resistance, using an adaptive mechanism which can be a proportional-integral (PI) or integral controller. These methods essentially utilize the machine model and its accuracy is therefore heavily dependent on the accuracy of the model used. In general these methods primarily differ with respect to which quantity is selected for adaptation purposes. A method which uses reactive power is not dependent on stator resistance at all and is probably the most frequently applied approach [20]-[22]. Other possibilities are to use torque as reported in [22], [23] or rotor flux magnitude as reported in [23] air gap power [24], *d*-axis stator voltage or *q*-axis stator voltage [25], stator fundamental rms voltage [26], rotor back emf [27] etc. One of the common features that all of the methods of this group share is that rotor resistance adaptation is usually operational in steady-states only and is disabled during transients.

2.3.4 Heuristic methods

There exist a number of other possibilities for online rotor resistance adaptation which can not be classified in the methods described in sections 2.3.1, 2.3.2 and 2.3.3. Chan and Wang [28] have presented a new method for rotor resistance identification, with a new coordinate axes selection. They set a new reference frame which was coincident with the stator current vector. They measured the steady-state stator voltage, current and motor speed, and obtained the stationary reference frame components by using a three-phase to two phase transformation. The rotor resistance was then calculated algebraically with the equation they derived. Theoretically, this identification method is valid only for steady-state operation of the motor.

Toliyat *et al* [29] proposed a rotor time constant updating scheme, which neither required any special test signal nor any complex computation. This technique utilized a modified switching technique for the current-regulated pulse width modulation voltage source inverter to measure the induced voltage across the stator terminals. The induced voltage was measured at every zero crossing of the phase currents. Thus, for the three phase induction motor, the proposed technique provided six instants to update the rotor time constant. The technique was capable of measuring the rotor time constant for the minimum stator frequency of 5 Hz.

Another possibility, initiated by the recent developments in artificial intelligence, is the application of artificial neural networks and fuzzy logic for the on-line rotor time

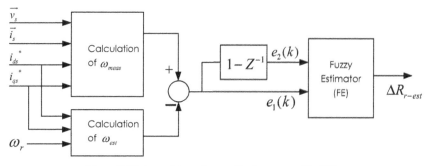

Figure 2.4 Rotor resistance fuzzy estimator proposed in [30].

constant / rotor resistance adaptation. In [30]-[32], there were attempts to use fuzzy logic; however none of them were supported by experimental data substantiating their modeling work. In the fuzzy rotor resistance updating scheme proposed by Zidani *et al* [30], a direct estimate of the stator frequency ω_{meas} is derived which is independent of rotor resistance, and is taken as the measured stator frequency as shown in Figure 2.4. On the other hand, the stator frequency implemented in the control system is taken as ω_{est}. The difference between these two estimates was used to map the rotor resistance with a fuzzy logic function. In this paper they have activated the fuzzy estimator only in the steady-state. Bim [31] proposed a fuzzy rotor time constant identification based on a fuzzy optimisation problem in which the objective function is the total square error between the commanded stator currents and measured stator currents in the *d-q* reference frame as indicated in Figure 2.5. Because the variation

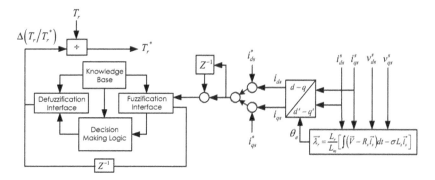

Figure 2.5 Fuzzy logic based T_r updating scheme for indirect FOC in [31].

of the motor thermal time constant is very slow compared with the motor electrical time constant, a sampling interval of 5 seconds was chosen. Inadequate modeling and experimental results were included. Ta-Cao *et al* [32] estimated the rotor resistance, with only the steady-state measurements assuming the resistance variation is very slow.

Their estimation was based on a characteristic function F defined by:

$$F = \frac{1}{\omega_e}\left(i_{qs} \frac{d\lambda_{dr}^s}{dt} - i_{ds} \frac{d\lambda_{qr}^s}{dt} \right) \qquad (2.28)$$

This characteristic function was estimated using the reference values as F_{est} and was calculated from the measured voltages and currents as F_{act}. The error between them is used to estimate the rotor resistance variation as shown in Figure 2.6.

Also Fodor *et al* [33], Ba-Razzouk [34] and Mayaleh *et al* [35] have reported that they adapted the rotor time constant using the artificial neural networks. Fodor *et al* [33] investigated the possibility of using a neural network to compensate parameter variations in the IFO controlled induction motor drive. They added an off-line trained neural network as a black box, which estimates T_r to the IFO controller as shown in Figure 2.7. The four inputs to the neural network are $i_{ds}, i_{qs}, i_{ds}^*, i_{qs}^*$, the measured and

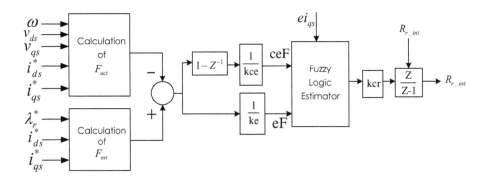

Figure 2.6 Rotor resistance estimator using fuzzy logic proposed in [32].

commanded stator currents in the *d-q* rotating reference frame. A rotor flux observer has been used to estimate the flux angle used for *d-q* transformation. The neural network was trained with a specific I/O pattern selected by the authors and only the steady-state model of the IFO controller was used. The major drawback of this type of learning is that the network may calculate the output with very large errors if the drive has to operate through an I/O pattern which was not envisaged in the learning process. This work was also not supported by experimental results of the implementation.

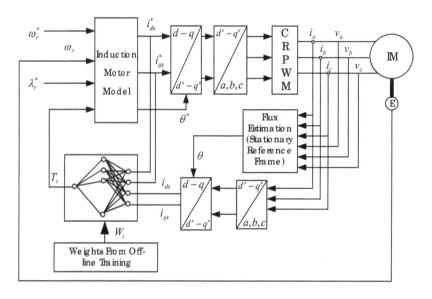

Figure 2.7 IFO control block with T_r adaptation using neural networks in [33].

Ba-Razzouk *et al* [34] proposed another ANN method for rotor time constant adaptation in IFO controlled drives. There are five inputs to the T_r estimator using neural network, namely $v_{ds}^s, v_{qs}^s, i_{ds}^s, i_{qs}^s, \omega_r$. The training signals are generated with step variations in rotor resistance for different torque reference T_e^* and flux command λ_r^* and the final network is connected in the IFO controller as shown in Figure 2.8. The rotor time constant was tracked by a PI regulator that corrects any errors in the slip calculator. The output of this regulator is summed with that of the slip calculator and the result constitutes the new slip command that is required to compensate for the rotor time constant variation. The major drawback of this scheme is that the final neural network is only an off-line trained network with a limited data file in the modeling.

Mayaleh *et al* [35] proposed a rotor time constant estimation using a recurrent neural network. Their algorithm used the three stator voltage and three stator current measurements in the stator reference frame. The rotor time constant was obtained at the output of a recurrent neural network (RNN) as shown in Figure 2.9. The three inputs to the RNN were stator currents, rotor fluxes and rotor speed. The rotor flux was calculated using motor parameters and the influence of stator resistance on rotor flux estimation was not accounted for. Even though the results and the method employed were elegant, these results were not backed up by experimental data subsequently.

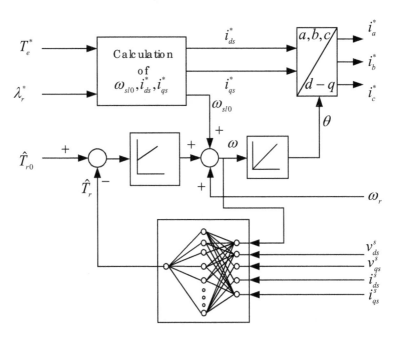

Figure 2.8 Principle of rotor time constant adaptation in [34].

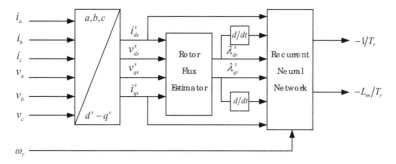

Figure 2.9 Rotor time constant estimator proposed by Mayaleh *et al* [35].

2.4 On-line stator resistance estimation techniques

In section 2.3, only the rotor resistance estimation techniques were reviewed for RFOC implementations where a speed sensor was mounted on the induction motor shaft. IFO schemes which do not require the stator resistance for satisfactory operation have become the industrially accepted standard for speed sensor based rotor flux orientated control. The situation has however dramatically changed with the advent of sensorless RFOC, which requires estimation of rotor speed.

The rotor speed of an induction motor can be synthesized from its state equations, and can be written as:

$$\omega_r^{est} = \frac{1}{\lambda_r^2}\left[\left(\lambda_{dr}^{s\,vm}\frac{d\lambda_{qr}^{s\,vm}}{dt} - \lambda_{qr}^{s\,vm}\frac{d\lambda_{dr}^{s\,vm}}{dt}\right) - \frac{L_m}{T_r}\left(\lambda_{dr}^{s\,vm}i_{qs}^s - \lambda_{qr}^{s\,vm}i_{ds}^s\right)\right] \qquad (2.29)$$

The *d*- axis and *q*-axis rotor flux linkages used in eqn (2.29) can be estimated using equations (2.10) - (2.13). As the stator resistance R_s is used in these equations, the accuracy of this type of speed estimation depends on the value of stator resistance. The detuning of the stator resistance leads to large speed estimation errors and could even lead to instability at very low speeds. It is for this reason that on-line estimation of stator resistance has received considerable attention during the last decade, as witnessed by a large number of publications [36]-[44] which address this issue. In this section, some of the stator resistance identification schemes reported in the literature is reviewed briefly.

The method proposed by Kerkman *et al* [36] used a back electromotive force (BEMF) detector- which decouples direct component of stator flux from the corrupting effects of the stator resistance, producing an ideal signal for adaptation algorithms. In addition, the quadrature component provides a near instantaneous estimate of the stator resistance as shown in Figure 2.10.

Habetler *et al* [37] proposed an instantaneous hybrid flux estimator to tune both the stator and rotor resistances for stator flux oriented induction motor drives used in electric vehicles operating over a wide speed range. The stator flux was determined accurately from the terminal voltages when the motor was operating at higher speed. However, at low speed, the rotor flux was found using the motor speed and rotor time constant (current model) and the stator flux was estimated from the rotor flux. By

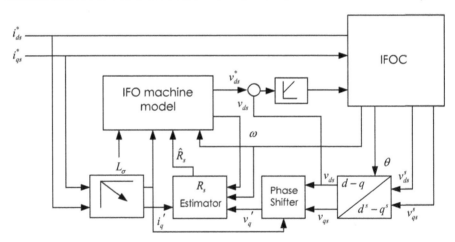

Figure 2.10 Stator resistance identifier using BEMF detector by Kerkman *et al* [36].

alternating between these two methods of determining stator flux, a self tuning operation was achieved. The rotor resistance was estimated when the stator flux calculation was carried out with a voltage model. Even though this method was implemented for a stator flux oriented induction motor drive this method was discussed here as the approach could be applicable to RFOC as well.

Marino *et al* [38] addressed the problem of simultaneous on-line estimation of both rotor and stator resistances based on the measurements of rotor speed, stator currents and stator voltages. Their main contribution was in designing a novel ninth order estimation algorithm which contains both rotor flux and stator current estimates. Their design goal was to force stator current estimation errors to tend asymptotically to zero for any initial condition. They have shown in this paper that both stator and rotor resistance estimates converged exponentially to the true values for any unknown value of stator and rotor resistances. The experiments were conducted only to test the performance of the identification algorithm in the presence of sensor noise, inverter voltage distortion, quantization effect and modelling inaccuracies and not in the context of improving the performance of any induction motor controller scheme.

Bose *et al* described in [39], a quasi-fuzzy method of on-line stator resistance estimation of an induction motor, where the resistance value is derived from stator winding temperature estimation as a function of stator current and frequency through an approximate dynamic thermal model of the machine. The dynamic thermal model of the machine can be approximately represented by a first order low-pass filter as indicated in Figure 2.11. Once the steady-state temperature is estimated by the fuzzy estimator block, it is then converted to dynamic temperature rise through the low-pass

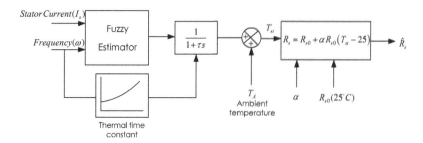

Figure 2.11 Quasi-fuzzy stator resistance estimation block diagram by Bose *et al* in [39].

filter and added to ambient temperature T_A to derive the actual stator temperature T_{st}. Neglecting the small amount of skin and stray loss effects, the stator resistance $\hat{R}_s$ is then estimated from the measured temperature rise of the stator winding using the equation shown in Figure 2.12.

Guidi *et al* proposed [40] another stator resistance estimation method for a speed sensorless DFO using a full order observer. Here the stator resistance estimation is based on a two-time scale approach, and the error between the measured and observed current is used for parameter tuning. The stator resistance adaptive law is depicted in Figure 2.12. The measured and observed stator currents are transformed to a reference frame whose x-axis is oriented with the stator voltage vector v_s and the stator resistance is estimated using the adaptive law proposed. If the speed estimation loop is much faster than the resistance estimator, the stability of the resistance estimator is not influenced. In order to get a reasonable performance, the speed adaptive law must ensure speed estimation dynamics at least comparable to the mechanical dynamics of the drive. To satisfy the above condition the resistance estimation is run very slowly.

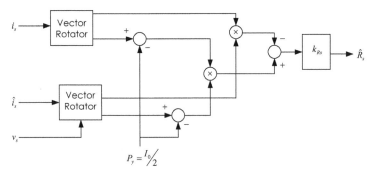

Figure 2.12 Stator resistance estimator proposed by Guidi in [40].

Tajima, Guidi *et al* [41] modified the adaptive observer proposed in [40] with an additional rotor resistance estimator by adding an additional harmonic to the *d*-axis current command. The injected frequency was kept low enough so that the related skin effect is negligible. Because of the unique relationship between the magnitude of the harmonic *d*-axis current and the rotor resistance, the latter was estimated using another adaptive law.

At present, speed sensorless RFOC exhibits poor dynamic and steady-state performance at very low speeds because of the following reasons:

- low offset and drift components in the acquired feedback signals
- voltage distortions caused by non-linear behavior of the switching converter
- increased sensitivity against model parameter mismatch

Holtz *et al* [42] proposed a scheme to compensate the above three drawbacks. They used a pure integrator for stator flux estimation which permits higher estimation bandwidth. Increased accuracy was achieved by eliminating direct stator voltage measurement, instead, the reference voltage corrected by a self-adjusting non-linear inverter model was used. The time-varying disturbances were compensated by an estimated offset voltage vector. The stator resistance estimation algorithm relies on the orthogonal relationship between the stator flux vector and the induced voltage in the steady-state. The stator resistance was found from the inner product of these two vectors in current coordinates. Current coordinates is a reference frame aligned with the current vector. The signal flow graph of this stator resistance estimation scheme is shown in Figure 2.13. Here, γ is the angle of stator current vector from the stator coordinates and δ is the angle of stator flux from the stator coordinates.

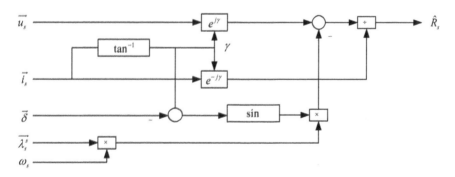

Figure 2.13 Signal flow graph of the stator resistance estimator by Holtz and Quan [42].

For speed sensorless operation of induction motor, Akatsu and Kawamura proposed an R_r estimation algorithm without the need for an additional signal injection supported by a new R_s estimation as discussed in [43]. The stator resistance R_s was estimated such that the rotor flux reference $\overrightarrow{\lambda_r^*}$ and the estimated rotor flux $\overrightarrow{\lambda_r}$ become the same because the error $\overrightarrow{\lambda_r^*} - \overrightarrow{\lambda_r}$ is in proportion to the estimation error of R_s. The block diagram for R_s estimation is shown in Figure 2.14. The simultaneous compensation of R_s and R_r was made possible only with the help of an ideal power op-amp instead of a voltage source inverter and hence its practical application is not expected for industrial drives.

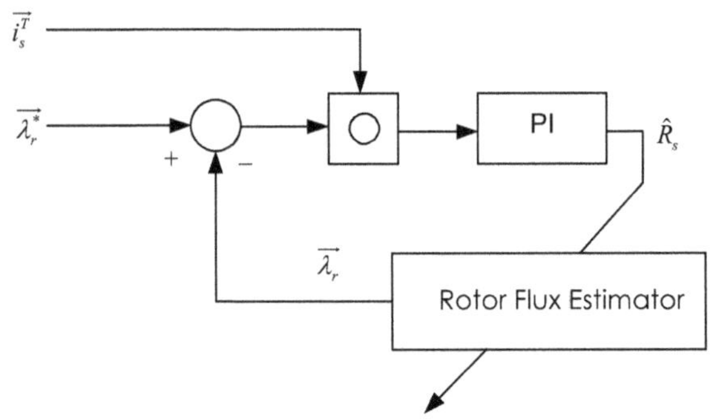

Figure 2.14 Block diagram of R_s estimation by Akastsu [43].

Ha and Lee [44] proposed another identification algorithm for stator resistance which has been based on the steady-state power flow between stator and rotor through the air gap. The steady-state power across the air gap was represented as the difference between the steady-state input power and the steady-state power dissipated by the stator windings. The air gap power was calculated using the steady-state value of estimated torque. The stator resistance was then estimated using the difference between these two steady-state powers. They have reported that the identification algorithm should be executed only in the steady-state and cannot do estimations during transient conditions.

The basic principles underlying field orientation by the two methods namely: direct and indirect field orientation and their influence on motor parameters are discussed in section 2.2. Most of the techniques for rotor resistance/rotor time constant identification reported in the literature are briefly reviewed in section 2.3. Also, there have been attempts to identify stator resistance of an induction motor either because they used stator flux orientation or the drive is speed sensorless. Some of these techniques are reviewed in section 2.4.

2.5   Conclusion

Artificial Neural networks have the advantages of extremely fast parallel computation, immunity from input harmonic ripple and fault tolerance characteristics due to the distributed network. Also, in recent years, fuzzy logic has emerged as an important Artificial Intelligence tool to characterize and control a system whose model is not known or not well defined. The next chapter describes how an Artificial Neural Networks and fuzzy logic could enhance the estimator and some of their applications in induction motor drives.

CHAPTER 3

REVIEW OF APPLICATION OF ARTIFICIAL NEURAL NETWORKS AND FUZZY LOGIC FOR ESTIMATION IN INDUCTION MOTOR DRIVES

3.1 Introduction

In classical control systems, knowledge of the controlled system (plant) is required in the form of a set of algebraic and differential equations, which analytically relate inputs and outputs. However, these models can become complex, rely on many assumptions, may contain parameters which are difficult to measure or may change significantly during operation as in the case of the RFOC IM drive. Classical control theory suffers from some limitations due to the assumptions made for the control system such as linearity, time-invariance etc. These problems can be overcome by using artificial intelligence based control techniques, and these techniques can be used, even when the analytical models are not known. Such control systems can also be less sensitive to parameter variation than classical control systems.

Generally, the following two types of intelligence based systems are used for estimation and control of drives, namely:

a.   *Artificial Neural Networks (ANNs)*

b.   *Fuzzy Logic Systems (FLSs)*

The main advantages of using AI- based controllers and estimators are:

- Their design does not require a mathematical model of the plant.
- They can lead to improved performance, when properly tuned.
- They can be designed exclusively on the basis of linguistic information available from experts or by using clustering or other techniques.

- They may require less tuning effort than conventional controllers.
- They may be designed on the basis of data from a real system or a plant in the absence of necessary expert knowledge.
- They can be designed using a combination of linguistic and response based information.

The different types of function approximators such as the conventional, the fuzzy and neural function estimators and how these estimators are arrived at are discussed briefly in section 3.2. The neural estimators have been used in the estimation of flux, torque etc for induction motor drives. Some of these estimators are briefly reviewed in section 3.3. Also there have been some fuzzy estimators employed in induction motor drives, which are reviewed in section 3.4.

3.2   Conventional, fuzzy and neural function approximators

3.2.1 Conventional approximator

A conventional function approximator should give a good prediction, when a system is presented with new input data. A conventional function approximator, uses a mathematical model of the system as shown in Figure 3.1(a). Sometimes it is not possible to have an accurate mathematical model of the system, in such a case; mathematical model free approximators are required, as shown in Figure 3.1 (b) [45]. Artificial Neural Networks, Fuzzy logic based estimator and fuzzy-neuro networks are function approximators, which can replace a conventional function approximator.

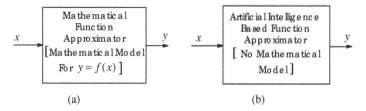

(a)   (b)

Figure 3.1 Conventional and artificial-intelligence based function approximators.

3.2.2 Neural function approximator

The function $y = f(x_1, x_2, ..., x_n)$, is the function to be approximated and x_1, x_2, x_3,....x_n are the n variables (n inputs) and the approximator uses the sum of non-linear functions g_1, g_2, g_3, g_4,....g_i , where each of the g are non-linear functions of a single variable.

$$y = f(x_1, x_2, ..., x_n) = g_1 + g_2 + g_3 + ... + g_{2n+1} = \sum_{i=1}^{2n+1} g_i \qquad (3.1)$$

where g_i is a real and continuous non-linear function which depends only on a single variable z_i.

$$g_i = g_i (z_i) \qquad (3.2)$$

$$z_i = h_{ji}(x_j) = h_{1i}(x_1) + h_{2i}(x_2) + ... h_{ni}(x_n) \qquad (3.3)$$

Thus it follows

$$g_1 = g_1(z_1) = g_1(h_{11}(x_1) + h_{21}(x_2) + ... + h_{n1}(x_n)) \qquad (3.4)$$

$$g_2 = g_2(z_1) = g_2(h_{12}(x_1) + h_{22}(x_2) + ... + h_{n2}(x_n)) \qquad (3.5)$$

•

•

•

$$g_{2n+1} = g_{2n+1}(z_{2n+1}) = g_{2n+1}(h_{12n+1}(x_1) + h_{22n+1}(x_2) + ... + h_{n2n+1}(x_n)) \qquad (3.6)$$

Thus in a very compact way, the non-linear function approximator can be mathematically described by (3.7).

$$y(x_1, x_2, x_3, ... x_n) = \sum_{i=1}^{2n+1} g_i \left\{ \sum_{j=1}^{n} h_{ji}(x_j) \right\} \qquad (3.7)$$

In the above equation, there are 2n+1 number of g_i functions. It is possible to make a modification, where h_{ji} is replaced by $\lambda_i h_j$, where λ_i are constants and h_j are strictly increasing continuous functions. Thus the function approximator can be defined as follows:

$$y(x_1, x_2, x_3, ...x_n) = \sum_{i=1}^{2n+1} g_i \left\{ \sum_{j=1}^{n} h_{ji}(x_j) \right\} = \sum_{i=1}^{2n+1} g_i \left\{ \lambda_i \sum_{j=1}^{n} h_j(x_j) \right\} \qquad (3.8)$$

Let g be a non-constant, bounded and monotonic increasing continuous function, $g(S_j) = 1 / [1 + \exp(-S_j)]$. Given that $y = f(x_1, x_2, x_3, \ldots x_n)$, and if $h_j(x_j) = S_j = w_{ij}x_j - \theta_1$, where w_{ij} and θ_i are real constants ($i = 1, 2 \ldots M = 2n+1, j = 1, 2, \ldots n$).

It follows that

$$y(x_1, x_2, x_3, \ldots x_n) = \sum_{i=1}^{M} \lambda_i g_i \left\{ \sum_{j=1}^{n} w_{ij}x_j - \theta_i \right\} \tag{3.9}$$

If for simplicity, $\theta_1 = 0$, and $g_1 = g_2 = \ldots = g$, then

$$y(x_1, x_2, x_3, \ldots x_n) = \sum_{i=1}^{M} \lambda_i g_i \left\{ \sum_{j=1}^{n} w_{ij}x_j \right\} = \lambda_{1g}(s_1) + \lambda_{2g}(s_2) + \ldots + \lambda_{Mg}(s_M) \tag{3.10}$$

where,

$$S_1 = \sum_{j=1}^{n} w_{1j}x_j = w_{11}x_1 + w_{12}x_2 + \ldots w_{1n}x_n \tag{3.11}$$

$$S_2 = \sum_{j=1}^{n} w_{2j}x_j = w_{21}x_1 + w_{22}x_2 + \ldots w_{2n}x_n \tag{3.12}$$

•

•

$$S_M = \sum_{j=1}^{n} w_{Mj}x_j = w_{M1}x_1 + w_{M2}x_2 + \ldots w_{Mn}x_n \tag{3.13}$$

These equations can be represented by a network shown in Figure 3.2. There are n input nodes in the input layer, M hidden nodes in the hidden layer.

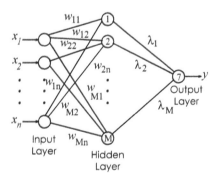

Figure 3.2 Neural network to yield the required function.

Figure 3.3 shows the schematic diagram of a neural approximation and Figure 3.4 shows the technique of its training. The error (*e*) between the desired non-linear function (*y*) and the non-linear function (*ŷ*) obtained by the neural estimator is the input to the learning algorithm for the artificial neural estimator. This type of learning is known as back propagation.

Figure 3.3 Schematic of a neural function approximator.

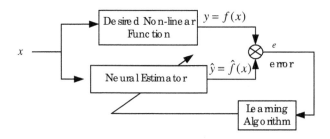

Figure 3.4 Training of a neural function approximator.

The output of a single neuron can be represented as

$$a_i = f_i \left\{ \sum_{j=1}^{n} w_{ij} x_j(t) + b_i \right\}$$ (3.14)

where f_i is the activation function and b_i is the bias. Figure 3.5 shows a number of possible activation functions in a neuron. The simplest of all is the linear activation function, where the output varies linearly with the input but saturates at ±1 as shown with a large magnitude of the input. The most commonly used activation functions are nonlinear, continuously varying types between two asymptotic values 0 and 1 or -1 and +1. These are respectively the sigmoidal function also called log-sigmoid and the hyperbolic tan function also called tan-sigmoid. A reader with very little exposure to the fundamentals of ANN, could refer to introduction to the artificial neural networks in Appendix A.1.

36

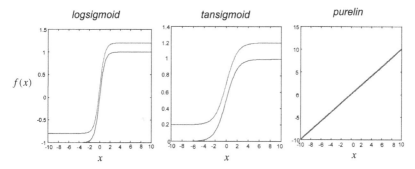

Figure 3.5 Logsigmoid, tansigmoid and purelin activation functions.

The learning process of an ANN is based on the training process. One of the most widely used training techniques is the error back propagation technique; a scheme which is illustrated in Figure 3.4. When this technique is employed, the ANN is provided with input and output training data and the ANN configures its weights. The training process is then followed by supplying with the real input data and the ANN then produces the required output data.

The total network error (sum of squared errors) can be expressed as

$$E = \frac{1}{2}\sum_{k=1}^{P}\sum_{j=1}^{K}\left(d_{kj} - o_{kj}\right)^{2} \tag{3.15}$$

where E is the total error, P is the number of patterns in the training data, k is the number of outputs in the network, d_{kj} is the target (desired) output for the pattern K and o_{kj} (= y_{kj}) is the jth output of the kth pattern. The minimization of the error can be arranged with different algorithms such as the gradient descent with momentum, Levenberg-Marquardt algorithm, reduced memory Levenberg-Marquardt algorithm, Bayesian regularization etc [92].

3.2.3 Fuzzy function approximator

A non-linear function can also be approximated by using a finite set of expert rules. These rules form a rule base and an individual rule can be considered as a rule of a sub expert. A fuzzy function approximator with a finite number of rules can approximate any continuous function to any degree of accuracy. Figure 3.6 shows the schematic of a fuzzy function approximator. Transformation 1 is the fuzzifier, because the crisp input data has to be transformed to linguistic values, as the rule base contains linguistic rules. Different types of membership functions used in transformation 1 are triangular, bell-shaped, sigmoid, gaussian, trapezoidal, S-shaped, T-shaped, L-shaped and so on. The fundamentals of fuzzy logic technology are described in Appendix A.2, to help readers who are not very familiar with the terminology in Fuzzy Logic.

The second block in Figure 3.6, the inference operation (decision making) is performed by using the rule base and knowledge base (which contains the knowledge of the system). The inference operation calculates the corresponding fuzzy (linguistic) output from the input. There is no mathematical equation or model used to describe the input-output relationship.

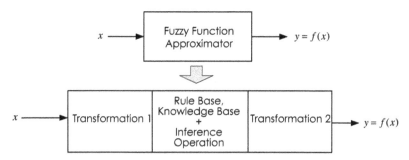

Figure 3.6 Fuzzy function approximator.

Similarly, transformation 2 transforms the linguistic output into crisp data and this is called the defuzzifier. Different types of membership functions used in transformation 1 are triangular, bell-shaped, sigmoid, gaussian, trapezoidal, S-shaped, T-shaped, L-shaped and so on. There are many defuzzification techniques used in transformation 2. The most widely used method is the centre of gravity (COG) method. For the continuous case the defuzzified output value is obtained from the overall membership function as follows:

$$Z^{*COG} = \frac{\int z\mu(z)dz}{\int \mu(z)dz} \qquad (3.16)$$

where $\mu(z)$ is the aggregate output membership function, z is the output quantity and $\int$ denotes algebraic integration. Some other methods used are the Height (H) method, mean - max method, first of maxima method, last of maxima and so on.

3.3 ANN based estimation in induction motor drives

Artificial neural networks have found widespread use in function approximation. It has been shown that, theoretically, a three layer ANN can approximate arbitrarily closely, any non-linear function, provided it is non-singular [47]. This property has been exploited by a few researchers working in the induction motor drives area. Bose has explained some of the major developments in this area in [48]. Some of the ANN based estimators reported in the literature for rotor flux, torque and rotor speed of induction motor drive are discussed in the following section.

3.3.1 Example of a speed estimation

In general, steady-state and transient analysis of induction motors is done using space vector theory, with the mathematical model having the parameters of the motor. To estimate the various machine quantities such as stator and rotor flux linkages, rotor speed, electromagnetic torque etc, the above mathematical model is normally used. However, these machine quantities could be estimated without the mathematical model by using an ANN. Here no assumptions have to be made about any type of non-linearity.

As an example, the rotor speed of an induction motor can be estimated from the direct and quadrature axis stator voltages and currents in the stationary reference frame, as shown in Figure 3.7. An 8 x 7 x 1 three layer feedforward ANN is used in this case. The network has 8 inputs, 1 output, and 7 hidden layers. The input nodes were selected as equal to the number of input signals and the output nodes as equal to the number of output signals. The number of hidden layer neurons is generally taken as the mean of the input and output nodes. In this speed estimator ANN, 7 hidden neurons were selected.

Here the results of an experiment conducted on a 3.6 kW, 4 pole, 50 Hz induction motor is explained. For training the ANN, the experimental waveforms of direct and quadrature axis voltages and currents when an unloaded induction motor was switched on to its rated voltage, are used. Initially, the ANN was trained with one data file which contains the stator voltages and stator currents captured when the 415 V supply was switched on directly to the induction motor stator terminals. The inputs to the neural network are the stator voltages and stator currents as indicated in Figure 3.7 and the output of the network is the measured speed. The network was trained using the *trainlm* function, the Levenberg-Marquardt backpropagation algorithm in matlab and the error plot obtained for this training is shown in Figure 3.8. The network has converged after 1102 *epochs* when the sum squared error has fallen below the set *goal* in the training. At the end of training, when the network has converged, the weights and biases of the ANN are obtained, which are used in the testing face. With this trained network, we can predict the speed for another data file, captured when the induction motor was switched on to a 415V supply at another point of time. Figures 3.9 and 3.10 show the results of prediction carried out for two different cases like this. The dotted graph in Figure 3.9 shows the actual speed of the motor, whereas the continuous graph is the estimated speed, which is the result of

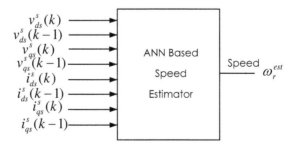

Figure 3.7 ANN based speed estimator for induction motor.

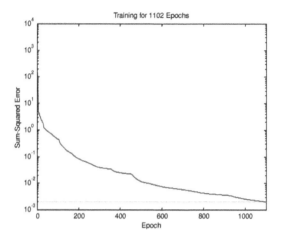

Figure 3.8 The Sum squared error plot when trained with 1000 data points.

prediction. The estimated speed is very close to the measured speed for case-1, where as they are not in close agreement for case-2. For training the ANN, only 1000 data pairs are used. To achieve a better prediction, the number of data pairs used for training has to be increased.

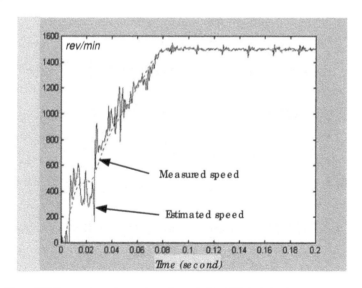

Figure 3.9 Measured speed vs. estimated speed with ANN for induction motor- case-1.

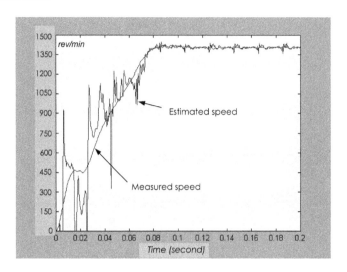

Figure 3.10 Measured speed vs. estimated speed by ANN for induction motor- case-2.

In order to investigate the case of better prediction by increasing the number of data pairs for training, a different experiment was conducted. A rotor flux oriented induction motor drive was set up in the laboratory, where the speed reference was changed in steps of 100 rpm and reversed every time the speed reached 1000 rpm. The load torque on the motor was kept constant at its full load rating. The stator voltages, stator currents and the rotor speed were measured for 5 seconds and a data file was generated. The neural network was then trained using the *trainlm* algorithm with this data file. The training has converged after 48 epochs and the error plot obtained for this network training is shown in Figure 3.11. Later, the estimated speed was predicted with the trained neural network, and the result is shown in Figure 3.12. The noisy data in the plot is the estimated speed and the continuous line is the speed measured with the encoder. The speed estimation was found to fail for speeds less than 100 rpm. If the trained neural network has to predict the speed under the complete range of operation of the drive, the data for training the neural network also has to be taken for the whole range.

From these two examples investigated, it was found that the off-line training of the neural network could not produce satisfactory results, and it can be concluded that these methods are not most suitable for these applications.

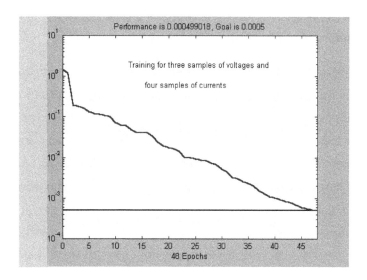

Figure 3.11 The error plot for higher number of samples.

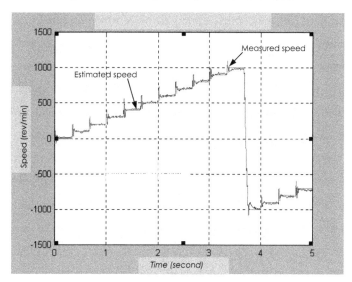

Figure 3.12 Measured speed vs. estimated speed with ANN for induction motor.

3.3.2 Flux and torque estimation

The same principle as described in Section 3.3.1 can also be extended for simultaneous estimation of more quantities such as torque and stator flux. When more quantities or variables have to be estimated, the complex ANN has to do a bigger non-linear mapping.

Toh *et al* [49] developed a rotor flux estimator for use in vector controlled induction motor drives using ANNs. Two ANNs are used, one for the magnitude of the rotor flux and the other for the sine of its space phase angle. The inputs to the ANNs are the present and the five previous samples of the d-q axis stator currents in the synchronously rotating reference frame. Standard multilayer ANNs with backpropogation are used. The maximum absolute error in the flux estimation was 0.03 pu for line-start operation of an induction motor. Even though Toh *et al* developed this estimator with an intention of using in a vector controlled induction motor drive, there was no effort reported in incorporating the above ANN based flux estimator into such a motor drive.

Mohamadian *et al* [50] have implemented an ANN which essentially computes the rotor flux angle and performs the transformation from the synchronously rotating reference frame to the stationary reference frame. The rotor flux angle is shown to depend on the present, the previous samples of the synchronously rotating d-q axis current and the rotor speed. The stationary d-q axis voltages and their previous samples are also given as inputs to the ANN to improve its accuracy. The ANN transforms the synchronous frame d-q axis current commands to the stationary frame d-q axis current commands. A 20-15-2 network is trained by the backpropagation algorithm to achieve the transformation. Although the authors used the terminology *ANN controller*, control action is not performed by their ANN. The neural network used for this transformation has 20 input neurons, 15 hidden neurons and two output neurons and it is only a feedforward network with time delay inputs. Its capability to estimate during dynamic conditions was not investigated.

In another study by Simoes and Bose [51], four feedback signals for a DFO induction motor drive have been estimated using ANNs. A 4-20-4 multilayer network has been used for the estimation of the rotor flux magnitude, the electromagnetic torque and the sine/cosine of the rotor flux angle. These authors have demonstrated both by modeling and experiments that the above estimated quantities were almost equal to the same quantities computed by a DSP based estimator. Both the estimated torque and rotor flux signals using neural network was found to have higher ripple content compared to the DSP based estimated quantities. It could be concluded that a properly trained ANN could totally eliminate the machine model equations as is evident from the results reported by these authors.

In one application, Ba-razzouk *et al* [52] have trained a 5-8-8-2 ANN to estimate the stator flux using the measured stator quantities. After training, the ANN is used in a DFO controlled drive. The rotor flux is computed from the stator flux estimate provided by the ANN and the stator current. This paper also presents an ANN based decoupler which is used for an IFO drive. A 2-8-8-1 ANN is used for implementing the mapping between the flux and torque references and the stator current references. The estimated rotor flux using an ANN and a conventional FOC controller was shown to be equal. These authors have used experimental data for the ANN training and thus effect of motor parameters was reduced. The authors were unable to use these estimated fluxes for controlling the induction motor in the experiment. The structure of the ANN used for this estimation is only that of a static ANN. It is preferable to have a dynamic neural network for this purpose.

Ben-Brahim *et al* [53] used an ANN based estimation for rotor speed together with the machine model, using linear neurons. Though the technique gives a fairly good estimate of the speed, this technique lies more in the adaptive control area than in neural networks. The speed is not obtained at the output of a neural network; instead, the magnitude of one of the weights corresponds to the speed. The four quadrant operation of the drive was not possible for speeds less than 500 rpm. The motor was not able to follow the speed reference during the reversal for speeds less than 500 rpm. The drive worked satisfactorily for speeds above 500 rpm. Even though this method does not fall into a true neural network estimator, the results achieved with this type of implementation were very good. The ANN implementation methods used in this thesis are very similar to the approach of these authors.

Kim *et al* [54] have developed a new ANN based speed estimator, where the speed is available at the output of a neural network. They have used a three layer neural network with five input nodes, one hidden layer and one output layer to give the estimated speed $\hat{\omega}_r(k)$ as shown in Figure 3.13. The three inputs to the ANN are a reference model flux λ_r^*, an adjustable model flux $\hat{\lambda}_r$ and $\hat{\omega}_r(k-1)$ the time delayed estimated speed. The multilayer and recurrent structure of the network makes it robust to parameter variations and system noise. The main advantage of their ANN structure lies in the fact that they have used a recurrent structure which is robust to parameter variations and system noise. These authors were able to achieve a speed control error of 0.6% for a reference speed of 10 rpm. The speed control error dropped to 0.584% for a reference speed of 1000 rpm.

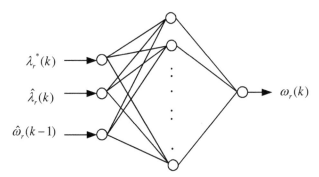

Figure 3.13 Structure of the neural network for the speed estimation in [54].

In addition to the estimators described in Section 3.3, some of the parameter identification schemes using ANN have already been discussed in Section 2.2.4.

3.4   Application of Fuzzy Logic techniques for estimation in induction motor drives

A fuzzy logic system essentially embeds the experience and intuition of a human plant operator, and sometimes those of a designer and/or a researcher of a plant. Fuzzy control is basically an adaptive and nonlinear control, which gives robust performance for a linear or a nonlinear plant with parameter variation. The advantages of using fuzzy logic based controllers in drives are:

- When parameters of the drive change (conventional controllers require significant redesign when plant characteristics, plant configuration and environment change),
- When existing controllers are to be augmented to self-tuning controllers,
- When it is easier to design and implement a fuzzy rule base because the plant is highly non-linear, when it is a complex environment that it is impossible to model efficiently.

As discussed in Section 3.1.2, the estimation capability of fuzzy logic was found to have applications such as slip gain tuning and efficiency optimisation in an IFO controlled induction motor drive. Sousa, Bose *et al* proposed a fuzzy MRAC based tuning controller in [55] for updating the slip gain in the IFO drive. The two inputs to the fuzzy tuning controller are one from the reference model and the other from the actual estimator, the slip gain was calculated at the output of the fuzzy tuning controller. In the ideally tuned condition of the system, both the reference model and the actual estimation signals will match, the error and change in error signal inputs will be zero and the slip gain will be set to the correct value. If the system becomes detuned, the actual signals will deviate from the respective reference values and the resulting error will alter the slip gain until the system becomes tuned. It can be noticed that the fuzzy logic based slip gain tuning system was able to adapt the slip gain satisfactorily in all the regions of the torque-speed plane.

In addition to the application as estimator, fuzzy logic has also been applied to the efficiency optimization problem. Sousa *et al* [56] have optimised the efficiency of an induction motor drive developed using a fuzzy efficiency controller. This method has used a fuzzy controller to adjust adaptively the magnetizing current based on the measured input power such that, for a given load torque and speed, the drive settles down to the minimum input power, thus achieving the optimum efficiency operation. The fuzzy controller enabled the fast convergence of the optimization problem.

Several applications of fuzzy logic techniques in parameter identification have already been discussed in Section 2.1.3.

3.5   Conclusion

The available literature on application of ANN and fuzzy techniques used for estimation in induction motor drives have been reviewed briefly in this chapter. These techniques have also found applications as controllers in these drives. However, these are not reviewed here, as the main focus of this thesis was only in the investigation of ANN and fuzzy techniques for rotor and stator resistance estimations in an induction motor drive. The problems encountered due to parameter detuning in the rotor flux oriented induction motor drive and an adaptation technique with a PI / fuzzy rotor resistance estimator is discussed in the next chapter.

CHAPTER 4

RFOC MODEL VALIDATION AND ROTOR RESISTANCE IDENTIFICATION USING PI AND FUZZY ESTIMATORS

4.1   Introduction

The detuning effect of an RFOC induction motor drive has been a major research topic over the last fifteen years. Many researchers have developed different solutions to this problem, which are reviewed in Chapter 2. The problem of detuning of an RFOC induction motor drive is revisited in this chapter. In order to study this problem in detail, a mathematical model of the drive has been developed and discussed in this section. The SIMULINK model developed has been validated with the help of an experimental set-up using a 1.1kW induction motor. The steady-state and dynamic performance of this drive has been presented. It has been noted that the performance results of the mathematical model developed are in very close agreement with the results obtained from the experimental set-up.

The effects of change in rotor resistance variation on the performance of the RFOC induction motor drive have been tested both in modelling and in the experimental set-up. The results of this investigation are presented in Section 4.2. A proportional integral (PI) estimator has been proposed for the adaptation of the rotor resistance in Section 4.3. This scheme has been analysed in detail using the SIMULINK models and the results are discussed in Section 4.3.1. The PI rotor resistance estimator has been implemented in the experimental set-up and the results are discussed in Section 4.3.2.

It has been noted in Chapter 3, that fuzzy logic could perform the role of a better non-linear mapping. For this reason, the problem of rotor resistance adaptation has been revisited with a fuzzy estimator in Section 4.4. This system was also analysed with the SIMULINK and fuzzyTECH models developed and the results are presented in

Section 4.4.1. It was found that developing the fuzzy estimator was very much simpler compared to the gain tuning of the PI estimator from the modeling studies.

4.2 Analysis and implementation of a RFOC induction motor drive

The main focus of this book was to look at the detuning effects of rotor resistance in a RFOC induction motor drive. It was thus necessary to develop both a mathematical model and a practical experimental set-up for this system. In this section, modeling and experimental results carried out for a 1.1kW induction motor drive are presented.

The mathematical background of indirect or feedforward vector control (referred here as RFOC) has already been discussed in Chapter 2. A block diagram of a conventional RFOC induction motor drive with PWM voltage source inverter is shown in Figure 4.1. The speed controller generates the input to the i_{qs} controller. The flux controller generates the reference to the i_{ds} controller. The currents i_{ds} and i_{qs} are controlled in the synchronously rotating reference frame. The decoupling unit removes the coupling caused by the i_d and i_q controllers. The flux model calculates

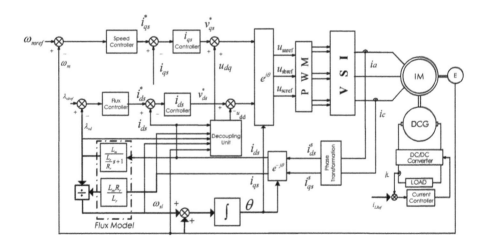

Figure 4.1 Block diagram - rotor flux oriented vector controlled induction motor drive.

the rotor flux based on the motor parameters. The estimated slip speed ω_{sl} is added to the measured rotor speed ω_r and integrated to get the position θ, which is then used for transformation of currents from the stationary reference frame to synchronously rotating reference frame and vice verse. The complete drive system has been modelled using SIMULINK [57]. All the modeling results obtained for this drive system have been validated with another set of results taken from the experimental set-up. The parameters of the induction motor used for this investigation are given in Table B.1.

4.2.1 Full-load torque step response

The drive has been implemented to run only the i_{qs} controller in Figure 4.1, to test this set of results by opening the speed controller loop. The modeling results have been shown in Figure 4.2. A step input of 7.4 Nm was applied to the motor shaft and

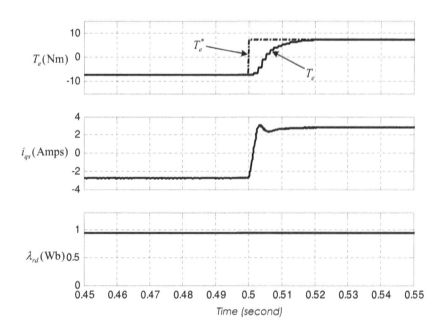

Figure 4.2 Full load torque step response - modeling results.

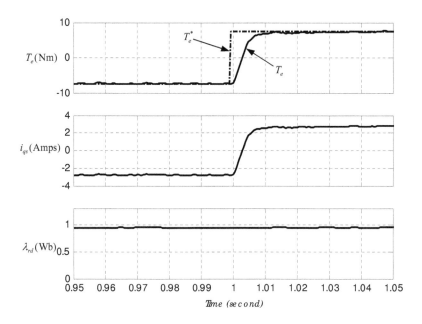

Figure 4.3 Full load torque step response - experimental results.

the step response has been recorded. The drive has taken close to 20 milliseconds to reach the steady-state in this case. The rotor flux amplitude and the quadrature axis stator current in rotating reference frame have also been recorded.

Later, the same torque step input of 7.4 Nm was applied in the experimental set-up, to obtain the equivalent figures from practical drive system. The results are shown in Figure 4.3. The step response time and the amplitudes of rotor flux and stator current are very close to the ones obtained in the modeling.

4.2.2 Full-load speed step response

The RFOC drive was also implemented incorporating the speed control loop of Figure 4.1. This test was modeled for the reversal of the drive at full load from 1000 rpm to   -1000 rpm. The modeling results are shown in Figure 4.4. In addition to the speed, the motor torque, amplitudes of the rotor flux linkage and quadrature axis stator current were recorded.

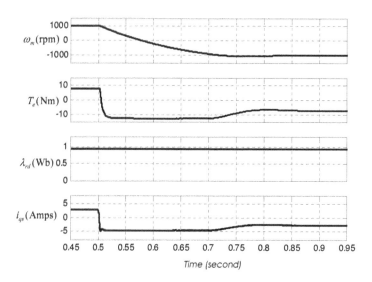

Figure 4.4 Step response for speed at full load – modeling results.

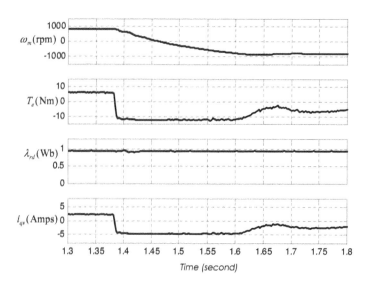

Figure 4.5 Step response for speed at full load – experimental results.

Later, the reversal of the motor was carried out in the experimental set-up for the same speed references, to obtain the equivalent figures from the practical drive system. The results are shown in Figure 4.5. The response time and the amplitudes of torque and currents are very close to the one obtained in the modeling.

In addition to this case, an additional experimental result for a step response from zero speed to 1000 rpm has been added, as shown in Figure 4.6, in order to demonstrate the drive's capability to run at zero speed.

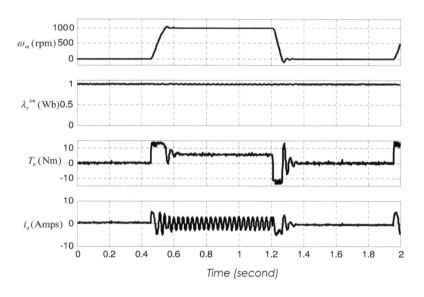

Figure 4.6 Step response from zero speed to 1000 rpm for induction motor with load – experimental result.

4.2.3 Full-load step response at fixed speed

Finally, a dynamic full load torque response test was conducted for the motor running at a fixed speed of 1000 rpm. The modelling results are shown in Figure 4.7. As the experimental set-up could produce only a delayed step torque, the modeling was conducted only with this load step profile. In addition to the speed, the motor torque, amplitudes of the rotor flux linkage and quadrature axis stator current have been recorded.

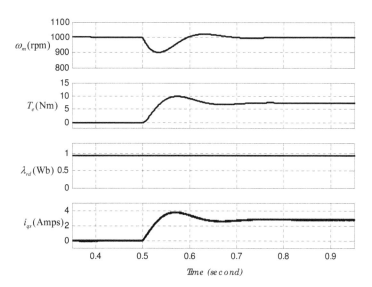

Figure 4.7 Full load step response at 1000 rpm – modeling results.

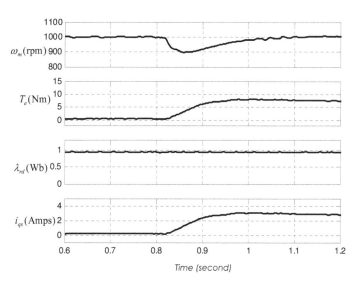

Figure 4.8 Full load step response at 1000 rpm – experimental results.

Later, the same load torque was applied to the induction motor shaft at the same speed of 1000 rpm, in the experimental set-up, to obtain the corresponding figures from the practical drive system. The results are shown in Figure 4.8. The response time and the amplitudes of torque and currents are very close to the one obtained in the modeling.

4.2.4 Experimental set-up for RFOC drive

An experimental set-up was built in the initial stages of the project and a photograph of the experimental set-up is shown in Figure E.1. The motor generator set consists of a 1.1kW three phase induction motor coupled to a 1.1kW permanent magnet DC motor. An encoder with 5000 pulses/rev was mounted to the induction motor shaft. A PC with Pentium III 800MHz was used to host the dSPACE DS1102 controller board and for the control software development. A functional block diagram is shown in Figure 4.9.

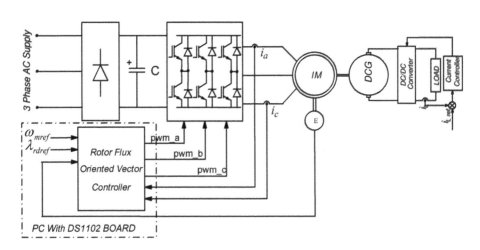

Figure 4.9 Block diagram of the experimental set-up for RFOC induction motor drive.

4.3 Effect of rotor resistance variation on the performance of the RFOC induction motor drive

The performance of the vector controlled drive depends on the accuracy of the estimated rotor flux from the measured stator currents, a mismatch between the real rotor flux and the estimated rotor flux, leads to error between the real motor torque and the estimated torque and results in poor dynamic performance. The accuracy of the estimated rotor flux is greatly determined by the accurate value of rotor resistance used for control. Rotor resistance may vary by up to 100% due to rotor heating and recovering this information with a temperature model or a temperature sensor is highly undesirable. In addition rotor resistance can change significantly with rotor frequency due to skew / proximity effect in machines with double-cage and deep-bar rotors. The problem related to rotor resistance adaptation has been reviewed extensively in [58].

The effect of temperature rise has been investigated on an experimental RFOC drive set up in the laboratory. Initially the parameters of the drive controller were initialized from the parameters obtained from the conventional no-load and blocked rotor tests described in Appendix-B. The drive was set up to run with a forward / reverses cycle at a frequency of around 0.4 Hz between ± 1000 rpm. When the motor was started, the temperature measured from a thermocouple inserted into the stator winding recorded 25°C. The rotor speed ω_m, estimated torque, T_e and the rotor flux linkage, λ_{rd} were recorded for this temperature and are shown in Figure 4.10. The motor was allowed to run for more than an hour and the stator winding temperature was recorded as 75°C. The parameters of the RFOC controller were kept unchanged during this experiment. The rotor speed, torque and rotor flux linkage were recorded for this temperature, as presented in Figure 4.11. It is obvious from Figure 4.11 that the drive has already started oscillating.

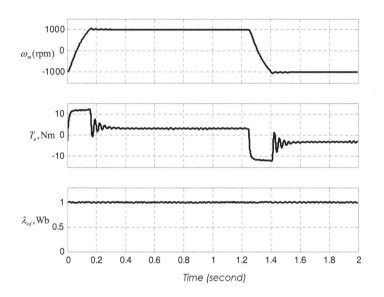

Figure 4.10 Experimental results- full load reversal with $R'_r = R_r$, stator temperature: 25°C.

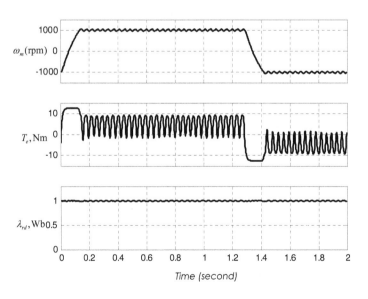

Figure 4.11 Experimental results- full load reversal with $R'_r = 0.611R_r$, stator temperature: 75°C.

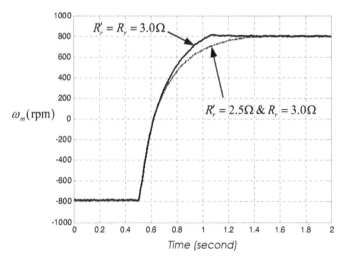

Figure 4.12 Effects of detuning of R_r : full load reversal- experimental results.

The effects of the detuned flux model in the RFOC induction motor drive was investigated again. The induction motor was allowed to run with a forward / reverses cycle at a frequency of around 0.2 Hz between ± 1000 rpm. The rotor resistance used in the controller R_r' was set to 3.0 Ω and the actual rotor resistance R_r was also 3.0 Ω. When these two resistances are equal, the correct acceleration profile of the motor is achieved, which has a higher slope as shown in Figure 4.12. However, when the rotor resistance in the controller R_r' was dropped to 2.5 Ω, then the acceleration of the motor was found to slow down, presenting a lower slope in this figure.

This chapter presents two methods of estimation of the rotor resistance in the RFOC induction motor drive to eliminate the problem observed in Figure 4.11. A model reference adaptive scheme has been proposed in which the adaptation mechanism is executed using either a PI estimator or a fuzzy estimator. The performance of both estimators and torque and flux responses of the drive are investigated with simulations for variations in the rotor resistance values from their nominal values. When either of the estimators is added to the drive system, the drive system performance does not deteriorate with the variation of the rotor resistance. The two estimation algorithms are designed, one with a PI controller and the other with fuzzy logic. The effectiveness of both algorithms has been demonstrated by simulations.

In this chapter, estimation of R_r is attempted with the error between the amplitude of the rotor flux computed by a current model $\left|\lambda_r^{im}\right|$ and voltage model $\left|\lambda_r^{vm}\right|$. Figure 4.13 shows the schematic diagram of the on-line rotor resistance tracking for the indirect vector controlled induction motor drive. Whenever there is variation in the real rotor resistance R_r of the motor, the rotor flux $\left|\lambda_r^{im}\right|$ estimated using the induction motor current model (Equation (4.2)), differs from the rotor flux $\left|\lambda_r^{vm}\right|$ estimated from the induction motor voltage model (Equation (4.1)). This error is then used to update the control rotor resistance R_r. The updating is done with both a PI controller estimator, as well with a fuzzy controller [59].

$$
\begin{bmatrix} \dfrac{d\lambda_{dr}^{s\,vm}}{dt} \\[2ex] \dfrac{d\lambda_{qr}^{s\,vm}}{dt} \end{bmatrix} = \frac{L_r}{L_m}\left\{ \begin{bmatrix} v_{ds}^s \\ v_{qs}^s \end{bmatrix} - R_s\begin{bmatrix} i_{ds}^s \\ i_{qs}^s \end{bmatrix} - \sigma L_s\begin{bmatrix} \dfrac{di_{ds}^s}{dt} \\[2ex] \dfrac{di_{qs}^s}{dt} \end{bmatrix}\right\}
\tag{4.1}
$$

$$
\begin{bmatrix} \dfrac{d\lambda_{dr}^{s\,im}}{dt} \\[2ex] \dfrac{d\lambda_{qr}^{s\,im}}{dt} \end{bmatrix} = \begin{bmatrix} -\dfrac{1}{T_r} & -\omega_r \\[2ex] \omega_r & -\dfrac{1}{T_r} \end{bmatrix}\begin{bmatrix} \lambda_{dr}^{s\,im} \\[2ex] \lambda_{qr}^{s\,im} \end{bmatrix} + \frac{L_m}{T_r}\begin{bmatrix} i_{ds}^s \\ i_{qs}^s \end{bmatrix}
\tag{4.2}
$$

The motor is controlled with the measured stator currents and the rotor position signal from the encoder using Equations (4.3) and (4.4) below.

$$
\lambda_{rd} = \frac{L_m}{1+pT_r}
\tag{4.3}
$$

$$
\omega_{slip} = \frac{L_m i_{qs}}{T_r \lambda_{rd}}
\tag{4.4}
$$

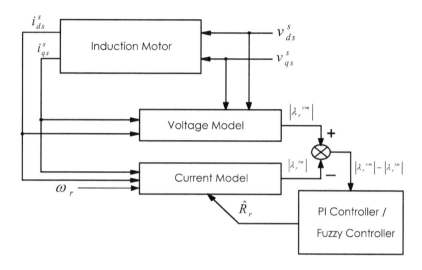

Figure 4.13 Structure of the model reference adaptive system for R_r estimation.

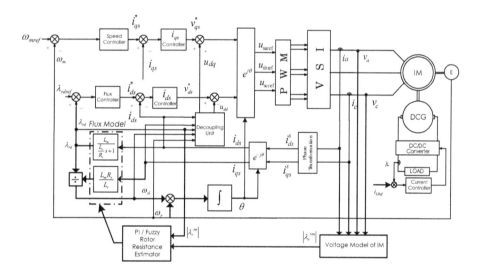

Figure 4.14 Block diagram of the RFOC induction motor drive with on-line PI / Fuzzy rotor resistance tracking.

In order to analyze the drive system performance for their flux and torque responses, with rotor resistance variation, the simulations were carried out for a 1.1kW (ABB) squirrel-cage induction motor. The PI/Fuzzy resistance estimator block in Figure 4.14 was kept disabled in the simulations, to study the effects of change in R_r.

Initially, the resistance R_r' used in the controller was kept at 6.03 Ω as shown Figure 4.15, and a step change of 40% was applied to the value R_r of the motor. The rotor flux linkage λ_{rd} increases to 1.2 Wb as shown in the same figure. It can also be seen that the steady-state value of the torque and i_{qs} the quadrature axis stator current in the rotating reference frame have reduced. As the variation of the rotor resistance of an induction motor is rather slow, a corresponding ramp change in R_r, has also been investigated and the results are shown in Figure 4.16. The rotor resistance change presented for the simulation was set to 40%. The simulation times are only to explain the concepts, the practical resistance variations depend on the thermal time constant of the motors.

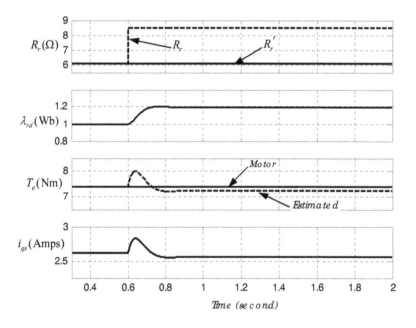

Figure 4.15 Effect of rotor resistance variation without R_r compensation for 40% step change in R_r - modeling results.

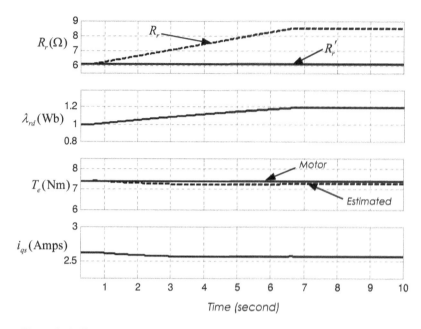

Figure 4.16 Effect of rotor resistance variation without R_r compensation for 40% ramp change in R_r – modeling results.

For the above cases, the motor was controlled at a constant speed of 1000 rpm with an Indirect Field Oriented Controller. A PWM switching frequency of 6.2 kHz was used for simulations to match with the experimental set-up.

4.4 Rotor resistance adaptation using a PI estimator

In the proposed PI rotor resistance estimator shown in Figure 4.17, the error between the $\left| \lambda_r^{im} \right|$ estimated by current model and the $\left| \lambda_r^{vm} \right|$ estimated by the voltage model is used to determine the incremental value of rotor resistance with a PI estimator and limiter. The incremental rotor resistance $\Delta \hat{R}_r$ is continuously added to the previously estimated rotor resistance $\hat{R}_{r0}$. The final estimated value $\hat{R}_r$ is obtained as the output of another low pass filter and limiter.

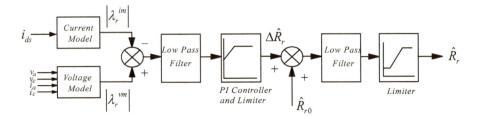

Figure 4.17 Block diagram of the PI rotor resistance estimator.

4.4.1 Modeling results using PI rotor resistance estimator

Dynamic simulations are performed using SIMULINK models for the drive, to validate the performance of the PI rotor resistance estimator. Fig 4.18 shows the results for a 40% step change in the rotor resistance for the 1.1kW ABB motor. The disturbances applied in this case are the same as for the uncompensated cases of

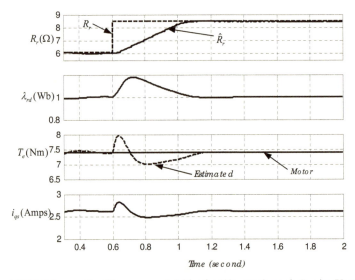

Figure 4.18 Modeling results of the compensated drive for 40% step change in R_r using PI estimator for R_r.

Figure 4.15. In the compensated case the estimated rotor resistance stabilizes to the actual resistance within around 300 milliseconds. Here, the sampling interval of 1 millisecond has been used for the resistance estimator. In the practical operating conditions, the rate of change of temperature is very slow and so is the change in rotor resistance. The simulation results presented in Figure 4.19 corresponds to this situation, where the rotor resistance variation is only of a slow ramp change. The uncompensated case for this situation is in Figure 4.16. Here the simulation was done only for 10 seconds and hence there are some changes in the estimated torque, because a higher slope is used for the rotor resistance variation. However, in practice the slope of the rotor resistance variation is very small and the error between real motor torque and the estimated one will be negligible.

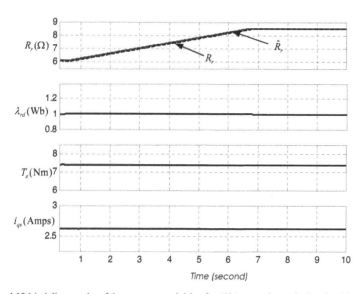

Figure 4.19 Modeling results of the compensated drive for 40% ramp change in R_r using PI estimator for R_r.

4.4.2 Experimental results using PI rotor resistance estimator

The rotor resistance estimator using a PI controller has been implemented in an RFOC induction motor drive set up in the laboratory. A block diagram for this implementation is shown in Figure 4.20. All the controllers required for this system have been implemented in software using a DS1102 dSPACE controller board. An IGBT inverter with a switching frequency of 5 kHz as shown in Appendix-D is used for the experiment. The sampling time used for the flux estimation, current and torque controllers is 300 μsecond, and it is 1200 μsecond for the speed controller. A separate 12 bit Data Acquisition Board AT-MIO-16E-10 manufactured by National Instruments together with Labview software is used, to log the data for 2-3 hours.

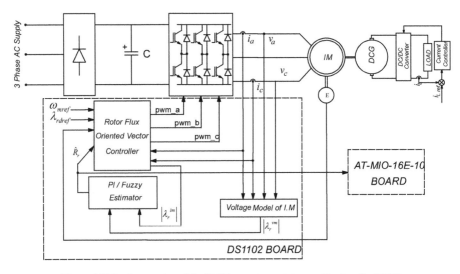

Figure 4.20 Implementation of the PI / Fuzzy rotor resistance estimators for RFOC.

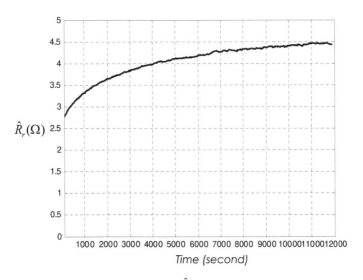

Figure 4.21 Estimated rotor resistance $\hat{R}_r$ using PI estimator – experimental results.

A heat run test was conducted with a load torque of 7.4 Nm, and the estimated rotor resistance was logged with the Labview software. Figure 4.21 shows the results obtained from this test.

4.5 Rotor resistance adaptation using a fuzzy estimator

In the proposed fuzzy logic estimator shown in Figure 4.22, the error between the $\left|\lambda_r^{im}\right|$ estimated by current model and $\left|\lambda_r^{vm}\right|$ estimated by the voltage model is used to determine the incremental value of rotor resistance through a fuzzy estimator. The incremental rotor resistance $\Delta \hat{R}_r$ is continuously added to the previously estimated rotor resistance $\hat{R}_{r0}$. The final estimated value $\hat{R}_r$ is obtained as the output of another low pass filter and limiter.

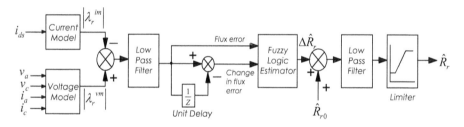

Figure 4.22 Block diagram of the fuzzy rotor resistance estimator.

The fuzzification stage input variables for the resistance estimator are *flux error*, *change in flux error*, and the output variable is *change in resistance*. The crisp input variables are converted into fuzzy variables using triangular membership functions as shown in Figures 4.23(a),(b). The rule base of the fuzzy logic estimator is shown in Figure 4.24. There are 49 rules (7x7), where *NL, NM, NS, Z, PS, PM, PL* correspond to *Negative Large, Negative Medium, Negative Small, Zero, Positive Small, Positive Medium, Positive Large* respectively.

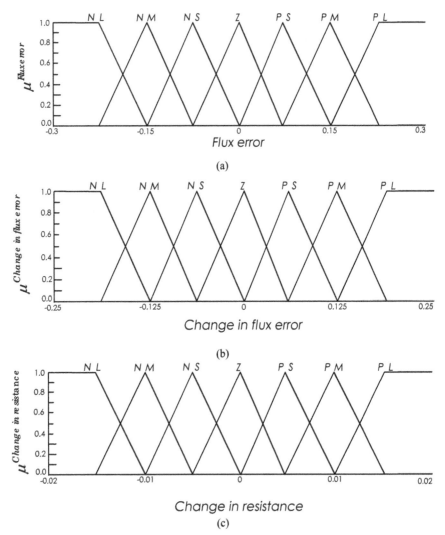

Figure 4.23 Membership functions for the fuzzy rotor resistance estimator.

In the defuzzification stage, a crisp value for the output variable, *change in resistance* ($\Delta \hat{R}_r$) is obtained by using the Mean of Maximum (MoM) operator [60]. The membership functions for the defuzzification stage are shown in Figures 4.23(c).

		Flux error						
		N L	N M	N S	Z	P S	P M	P L
Change In flux error	N L	N L	N L	N L	N L	N M	N S	Z
	N M	N L	N L	N L	N M	N S	Z	P S
	N S	N L	N L	N M	N S	Z	P S	P M
	Z	N L	N M	N S	Z	P S	P M	P L
	P S	N M	N S	Z	P S	P M	P L	P L
	P M	N S	Z	P S	P M	P L	P L	P L
	P L	Z	P S	P M	P L	P L	P L	P L

Figure 4.24 Rule base for the fuzzy rotor resistance estimator.

4.5.1 Modeling results of rotor resistance estimator using fuzzy controller

Dynamic simulations are repeated using SIMULINK models for the drive, to validate the performance of the fuzzy estimator. The fuzzy estimator is modeled using *fuzzy*TECH [61] and integrated into SIMULINK as an *s*-function. Figure 4.25 shows the results for a step change in the rotor resistance of the motor. The disturbances applied in this case are the same as for the uncompensated cases of Figure 4.15. In the compensated case the estimated resistance stabilizes to the actual resistance within nearly 300 milliseconds. The resistance estimator has a sampling interval of 1 millisecond. Figure 4.25 is a case where the resistance variation has a practical ramp change profile. The uncompensated results for a ramp change can be observed in Figure 4.16. Here the simulation has been done only for 10 seconds and hence there are some changes in the estimated torque, because a higher slope is used for the ramp change. The error between the estimated torque and the actual motor torque is negligible even for the 10 second simulation, but in the practical case the time for a

40% change could be of the order of one hour, resulting in a very good torque estimator.

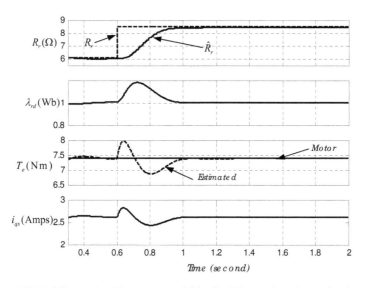

Figure 4.25 Modeling results of the compensated drive for 40% step change in R_r using fuzzy estimator.

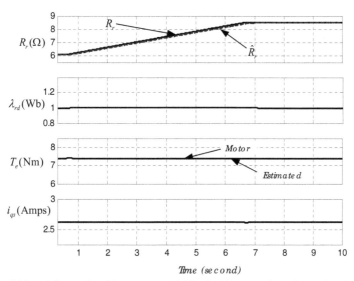

Figure 4.26 Modeling results of the compensated drive for 40% ramp change in R_r using fuzzy estimator.

4.6   Conclusions

In this chapter, the mathematical model of the RFOC induction motor drive was discussed and its validity was verified with the experiments carried out on a 1.1kW induction motor drive system. The effects on rotor resistance variations were investigated both in modeling and in the experimental set-up. Based on a model reference adaptive system, a rotor resistance identification scheme has been developed with the help of a simple PI estimator. The performance of this estimator was verified both in modeling studies and by experiment. Also, the fuzzy rotor resistance estimator was analyzed in detail using SIMULINK models.

It has been seen that tuning the gains of the PI estimator in the rotor resistance adaptation was rather difficult to avoid oscillations. It has been noted from the modeling results for load torques of 3.4 Nm and 7.4 Nm, shown in Figure 4.27 that the estimation time increased as the load torque reduced. However, when a fuzzy controller was used, the estimation times were not significantly different for the same load torques of 3.4 Nm and 7.4 Nm, shown in Figure 4.28.

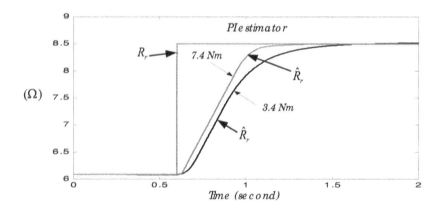

Figure 4.27 Modeling results of the compensated drive for 40% step change in R_r using PI estimator for R_r for two different load torques.

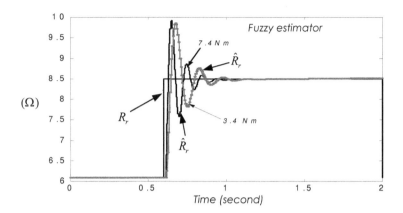

Figure 4.28 Modeling results of the compensated drive for 40% step change in R_r using fuzzy estimator for R_r for two different load torques.

The results exhibited by both PI and fuzzy rotor resistance estimators for both steady-state and transient conditions were compared and they were not significantly different. For this reason, the rotor resistance adaptation with fuzzy estimator was not carried out. The major problem with the PI estimator was in tuning the gains to avoid oscillation. Both of these estimators could not track the rotor resistance during dynamic conditions of the drive. To overcome the above problems, the adaptation approach implemented in this chapter was later improved with a new rotor resistance estimation using ANN, which is discussed in chapter 5.

CHAPTER 5

ROTOR RESISTANCE IDENTIFICATION USING ARTIFICIAL NEURAL NETWORKS

5.1 Introduction

Several techniques of identification of the rotor resistance for use in the RFOC drive have been discussed in Chapter 2. It was also shown in Chapter 3 that ANNs have the attributes of estimating parameters of non-linear systems with good accuracy and fast response. This chapter presents such an ANN based method of estimation for the rotor resistance of the induction motor in the RFOC induction motor drive. The backpropagation neural network technique is used for the real time adaptive estimation. The error between the desired state variable of an induction motor and the actual state variable of a neural model is back propagated to adjust the weights of the neural model, so that the actual state variable tracks the desired value. The performance of the neural estimator and torque and flux responses of the drive, together with this estimator, are investigated with simulations for variations in the rotor resistance from their nominal values.

The principle of on-line estimation of rotor resistance (R_r) with multilayer feedforward artificial neural networks using on-line training has been described in Section 5.3. This technique was then investigated with the help of modeling studies with a 3.6 kW slip-ring induction motor (SRIM), described in Section 5.4.1. In order to validate the modeling studies, modeling results were compared with those from an experimental set-up with a slip-ring induction motor under RFOC. These experimental results are presented in Section 5.4.2. As the main focus of this book is to find solutions for a squirrel-cage induction motor (SCIM), both modeling and experimental studies were also carried out for a 1.1 kW SCIM, discussed in Section 5.5.1 and Section 5.5.2 respectively. The slip-ring induction motors are used in industry only in very special applications because they are expensive and need maintenance because of their brushes. The majority of industrial drives use squirrel-cage induction motors. Hence, all further investigations were conducted only with squirrel-cage induction motors.

5.2 Multilayer feedforward ANN

Multilayer feedforward neural networks are regarded as universal approximations and have the capability to acquire nonlinear input-output relationships of a system by learning via the back-propagation algorithm as proposed by Funahashi [62] and by Hornik *et al* [63]. It should be possible that a simple two-layer feedforward neural network trained by the back-propagation technique can be employed in the rotor resistance identification. The modified technique using ANN proposed in this chapter can be implemented in real time so that the resistance updates are available instantaneously and there is no convergence issues related to the learning algorithm [64]. In this book, an alternative estimation of R_r is attempted with an Artificial Neural Network. The two-layered neural network based on a back-propagation technique is used to estimate the rotor resistance. Two models of the state variable estimation are used, one provides the actual induction motor output and the other one gives the neural model output. The total error between the desired and actual state

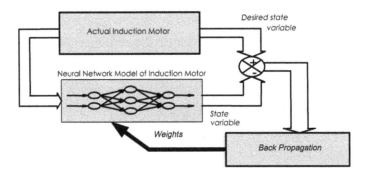

Figure 5.1 Block diagram of rotor resistance identification using neural networks.

variables is then back propagated as shown in Figure 5.1, to adjust the weights of the neural model, so that the output of this model coincides with the actual output. When the training is completed, the weights of the neural network should correspond to the parameters in the actual motor. Neural Networks have the ability to learn, so have become an attractive tool for process control.

5.3 Rotor resistance estimation for RFOC using ANN

The basic structure of an adaptive scheme described by Figure 5.1 is extended for rotor resistance estimation of an induction motor as illustrated in Figure 5.2. Two independent observers are used to estimate the rotor flux vectors of the induction motor. Equation (5.1) is based on stator voltages and currents, which is referred as the voltage model of the induction motor. Equation (5.2) is based on stator currents and rotor speed, which is referred as the current model of the induction motor.

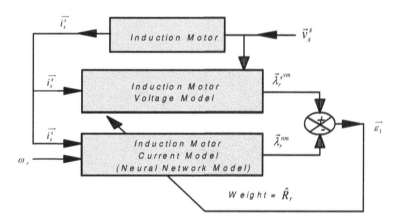

Figure 5.2 Structure of the neural network system for R_r estimation.

Voltage model equations:

$$\begin{bmatrix} \dfrac{d\lambda_{dr}^{s\,vm}}{dt} \\ \dfrac{d\lambda_{qr}^{s\,vm}}{dt} \end{bmatrix} = \frac{L_r}{L_m} \left\{ \begin{bmatrix} v_{ds}^s \\ v_{qs}^s \end{bmatrix} - R_s \begin{bmatrix} i_{ds}^s \\ i_{qs}^s \end{bmatrix} - \sigma L_s \begin{bmatrix} \dfrac{di_{ds}^s}{dt} \\ \dfrac{di_{qs}^s}{dt} \end{bmatrix} \right\} \tag{5.1}$$

Current model equations:

$$\begin{bmatrix} \dfrac{d\lambda_{dr}^{s\,im}}{dt} \\[4mm] \dfrac{d\lambda_{qr}^{s\,im}}{dt} \end{bmatrix} = \begin{bmatrix} -\dfrac{1}{T_r} & -\omega_r \\[4mm] \omega_r & -\dfrac{1}{T_r} \end{bmatrix} \begin{bmatrix} \lambda_{dr}^{s\,im} \\[4mm] \lambda_{qr}^{s\,im} \end{bmatrix} + \dfrac{L_m}{T_r}\begin{bmatrix} i_{ds}^{s} \\[2mm] i_{qs}^{s} \end{bmatrix} \tag{5.2}$$

The current model equation (5.2) can also be written as in (5.3).

$$\frac{d\overrightarrow{\lambda_r^{sim}}}{dt} = \left(\frac{-1}{T_r}I + \omega_r J\right)\overrightarrow{\lambda_r^{sim}} + \frac{L_m}{T_r}\overrightarrow{i_s^{s}} \tag{5.3}$$

where, $I = \begin{bmatrix} 1 & 0 \\ 0 & 1 \end{bmatrix}$, $J = \begin{bmatrix} 0 & -1 \\ 1 & 0 \end{bmatrix}$, $\overrightarrow{i_s^{s}} = \begin{bmatrix} i_{ds}^{s} \\ i_{qs}^{s} \end{bmatrix}$; $\overrightarrow{\lambda_r^{sim}} = \begin{bmatrix} \lambda_{dr}^{s\,im} \\ \lambda_{qr}^{s\,im} \end{bmatrix}$, $\overrightarrow{\lambda_r^{svm}} = \begin{bmatrix} \lambda_{dr}^{s\,vm} \\ \lambda_{qr}^{s\,vm} \end{bmatrix}$

The sample-data model of (5.3) is given in (5.4).

$$\overrightarrow{\lambda_r^{snm}}(k) = (W_1 I + W_2 J)\overrightarrow{\lambda_r^{snm}}(k-1) + W_3\overrightarrow{i_s^{s}}(k-1) \tag{5.4}$$

where, $W_1 = 1 - \dfrac{T_s}{T_r}$; $W_2 = \omega_r T_s$; $W_3 = \dfrac{L_m}{T_r} T_s$

Equation (5.4) can also be written as:

$$\overrightarrow{\lambda_r^{snm}}(k) = W_1 X_1 + W_2 X_2 + W_3 X_3 \tag{5.5}$$

where, $X_1 = \begin{bmatrix} \lambda_{dr}^{s\,nm}(k-1) \\ \lambda_{qr}^{s\,nm}(k-1) \end{bmatrix}$; $X_2 = \begin{bmatrix} -\lambda_{qr}^{s\,nm}(k-1) \\ \lambda_{dr}^{s\,nm}(k-1) \end{bmatrix}$; $X_3 = \begin{bmatrix} i_{ds}^{s}(k-1) \\ i_{qs}^{s}(k-1) \end{bmatrix}$

The neural network model represented by (5.5) is shown in Figure 5.3, where W_1, W_2 and W_3 represent the weights of the networks and X_1, X_2, X_3 are the three inputs to the network. If the network shown in Figure 5.3 is used to estimate R_r, W_2 is already known and W_1 and W_3 need to be updated.

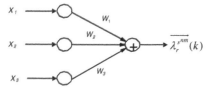

Figure 5.3 Two layered neural network model.

The weights of the network, W_1 and W_3 are found from training, so as to minimize the cumulative error function E_1,

$$E_1 = \frac{1}{2}\vec{\varepsilon}_1^{\,2}(k) = \frac{1}{2}\left\{\overline{\lambda_r^{svm}}(k) - \overline{\lambda_r^{sim}}(k)\right\}^2 \qquad (5.6)$$

The weight adjustment for W_1 using generalized delta rule is given by:

$$\Delta W_1(k) \propto -\frac{\partial E}{\partial W_1} = -\frac{\partial E}{\partial \overline{\lambda_r^s}(k)}\frac{\partial \overline{\lambda_r^s}(k)}{\partial W_1} \qquad (5.7)$$

$$-\frac{\partial E}{\partial W_1} = -\vec{\delta}X_1 \qquad (5.8)$$

where, $\quad \vec{\delta} = \dfrac{\partial E}{\partial \overline{\lambda_r^{snm}}(k)} = \left[\overline{\lambda_r^{snm}}(k) - \overline{\lambda_r^{snm}}(k)\right]^T$

$$\Delta W_1(k) = \left[\overline{\lambda_r^{snm}}(k) - \overline{\lambda_r^{snm}}(k)\right]^T I \overline{\lambda_r^{snm}}(k-1) \qquad (5.9)$$

The weight W_1 has to be updated using the current weight and the required correction as in (5.10).

$$W_1(k) = W_1(k-1) + \eta_1 \Delta W_1(k) \qquad (5.10)$$

where η_1 is the training coefficient.

To accelerate the convergence of the error back propagation learning algorithm, the current weight adjustment is supplemented with a fraction of the most recent weight adjustment, as in equation (5.11).

$$W_1(k) = W_1(k-1) - \eta_1\vec{\delta}X_2 + \alpha_1 \Delta W_1(k-1) \qquad (5.11)$$

where α_1 is a user-selected positive momentum constant.

Similarly the weight change for W_3 can be calculated as follows:

$$\Delta W_3(k) \infty - \frac{\partial E}{\partial W_3} = - \frac{\partial E}{\partial \overline{\lambda_r^s}(k)} \frac{\partial \overline{\lambda_r^s}(k)}{\partial W_3} \qquad (5.12)$$

$$-\frac{\partial E}{\partial W_3} = -\vec{\delta} X_3 \qquad (5.13)$$

$$W_3(k) = W_3(k-1) + \eta_1 \Delta W_3(k) \qquad (5.14)$$

As for W_1, in order to accelerate the convergence of the error back propagation learning algorithm, the current weight adjustments are supplemented with a fraction of the most recent weight adjustment, as in equation (5.15).

$$W_3(k) = W_3(k-1) - \eta_1 \vec{\delta} X_3 + \alpha_1 \Delta W_3(k-1) \qquad (5.15)$$

The rotor resistance R_r can now be calculated from either W_3 from (5.16) or W_1 from

(5.17), as follows:

$$\hat{R}_r = \frac{L_r W_3}{L_m T_s} \qquad (5.16)$$

$$\hat{R}_r = \frac{L_r (1-W_1)}{T_s} \qquad (5.17)$$

5.4 Modeling and experimental results with slip-ring induction motor

The focus of this book is on investigating the rotor resistance identification for squirrel- cage induction motors, which are widely used in the industry. However, the rotor resistance of a squirrel-cage induction motor cannot be measured by the conventional ammeter-voltmeter method used for measuring a resistance. Where as in the case of a slip-ring induction motor, the rotor resistance can be measured directly using the ammeter-voltmeter method. Also extra resistance of known value can be added into the rotor and then be compared with the estimated value using the rotor resistance estimation technique described in Section 5.3 for validation of the method.

For this reason, modeling and experimental investigations were carried out with a 3.6 kW slip-ring induction motor. The parameters of the slip-ring induction motor used are given in Table B.2 and the procedure for finding the parameters are described in Appendix-B.2.

5.4.1 Modeling results with slip-ring induction motor

Dynamic simulations were performed using SIMULINK models for the RFOC drive, to verify the approach of the proposed rotor resistance estimator technique described in Section 5.3. A schematic diagram showing the RFOC controller, rotor resistance estimator and the induction motor together with the loading arrangement is shown in Figure 5.4. The stator voltages and currents are measured to estimate the rotor flux linkages using the voltage model as shown in this figure. The inputs to the Rotor Resistance Estimator (RRE) are the stator currents, rotor flux linkages $\lambda_{dr}^{s\ vm}, \lambda_{qr}^{s\ vm}$ and the rotor speed ω_r. The estimated rotor resistance $\hat{R}_r$ will then be used in the RFOC controllers for the flux model.

The performance of the drive system was analyzed by adding an external resistance which is 40% of the original rotor resistance (R_r) into the rotor terminals, without altering the R_r' used in the RFOC controllers. The block - Rotor Resistance Estimator using ANN in Figure 5.4 was kept disabled in the simulations, to study the effects of change in R_r. The speed of the motor was kept constant at 1000 rpm. The changes in the estimated motor torque T_e, the rotor flux linkage λ_{rd}, and the current i_{qs} were noted. These results are shown in Figure 5.5. The estimated torque was found to reduce by nearly 5% and the estimated rotor flux linkage λ_{rd} was found to increase by nearly 21%. The rotor flux increase was only due to the fact that a linear magnetic circuit was assumed without magnetic saturation. The current i_{qs} was also found to follow a similar profile of the motor torque.

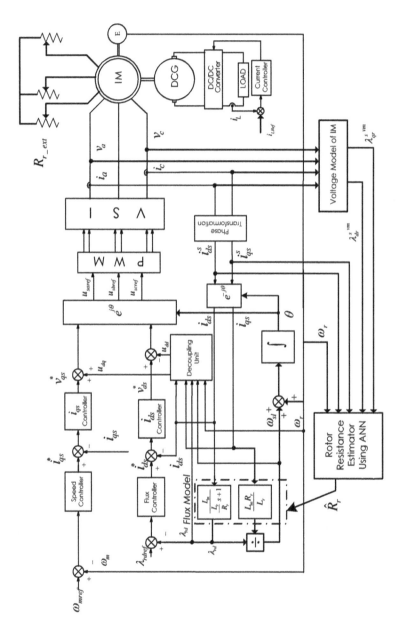

Figure 5.4 Block diagram of the RFOC slip-ring induction motor drive with ANN based on-line rotor resistance tracking.

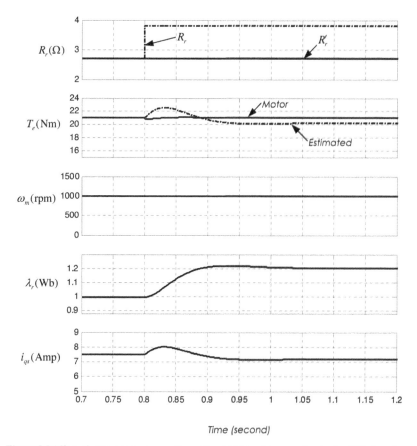

Figure 5.5 Effect of rotor resistance variation without rotor resistance estimator for 40% step change in R_r - modeling results (SRIM).

Subsequently simulation was repeated after enabling the RRE block, so that the rotor resistance in the controller R_r' was updated with the estimated rotor resistance $\hat{R}_r$. The results obtained for this case are presented in Figure 5.6. The estimated rotor resistance $\hat{R}_r$ has converged to the rotor resistance of the motor R_r within 50 milliseconds. The rotor resistance estimator has used a sampling interval of 100 microseconds. Now there is only a small transient disturbance in the rotor flux linkage λ_{rd} and estimated motor torque T_e.

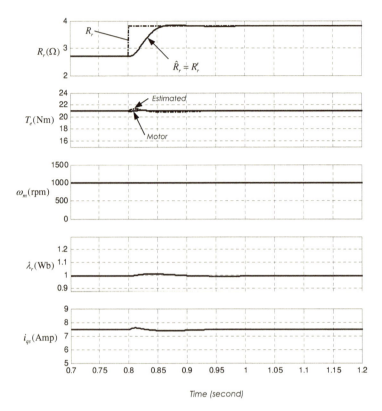

Figure 5.6 Effect of rotor resistance variation with rotor resistance estimator using ANN for 40% step change in R_r - modeling results (SRIM).

In the induction motor, the rate of temperature rise is very slow, because of the large thermal time constant. Consequently the change in rotor resistance due to temperature rise is only very slow. To investigate this situation, simulation was also carried out introducing a 40% ramp change in the rotor resistance over 8 seconds.

The performance of the drive system was analyzed by adding an external resistance with a value equivalent to 40% of the original rotor resistance into the rotor terminals, without altering R'_r the rotor resistance used in the RFOC controllers. The block - Rotor Resistance Estimator using ANN in Figure 5.4 was kept disabled in the simulations, to study the effects of change in R_r. The speed reference to the drive was kept constant at +1000 rpm for two seconds and then suddenly changed to -200 rpm. Subsequently the speed reference was taken to +400 rpm at 3 seconds, -600 rpm at 4 seconds, 700 rpm at 5 seconds, -800 rpm at 6 seconds and +1000 rpm at 7 seconds.

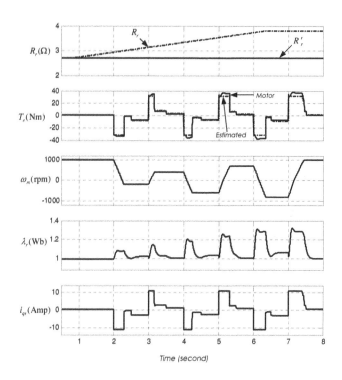

Figure 5.7 Effect of rotor resistance variation without Rotor Resistance Estimator for 40% ramp change in R_r - modeling results (SRIM).

Also the load torque was maintained at 1.0 Nm until 2.5 seconds, and then changed to 7.4 Nm at 2.5 seconds, 3.0 Nm at 3.5 seconds, 6.0 Nm at 4.5 seconds, 2.0 Nm at 5.5 seconds, 7.0 Nm at 6.5 seconds and 1.0 Nm at 7.5 second. The changes in the estimated motor torque T_e, the rotor flux linkage λ_{rd}, and the current i_{qs} were noted. These results are shown in Figure 5.7. The estimated torques were found to be different from the motor torques, as seen in the second trace from the top. The estimated rotor flux linkages were found to increase as the load torque was increased. The current i_{qs} followed the same profile as that of the torque.

Later, the simulation was repeated for the same speed references and load torques mentioned in the previous paragraph, after enabling the RRE block, so that the

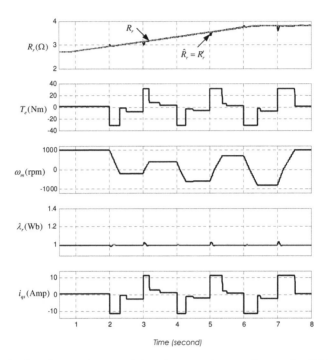

Figure 5.8 Effect of rotor resistance variation with rotor resistance estimator using ANN for 40% ramp change in R_r - modeling results (SRIM).

rotor resistance in the controller R'_r was updated with the estimated rotor resistance $\hat{R}_r$. The results obtained for this case, are presented in Figure 5.8. The estimated rotor resistance $\hat{R}_r$ was found to follow the rotor resistance of the motor R_r throughout. Only during the reversal of the motor, was a small error between R_r and $\hat{R}_r$ noticed. The rotor flux linkage λ_{rd} was found to remain almost constant during this simulation. The sampling interval of the rotor resistance estimator was 100 microseconds.

In order to show the difference between the actual and the estimated torques for the uncompensated and compensated cases described by Figures 5.7 and 5.8, they are plotted in a separate enlarged Figure 5.9. In the uncompensated case, Figure 5.9(a), there is an error between estimated torque and the motor torque in the steady-state. However, when the ANN based rotor resistance was switched on, the estimated and the real motor torques followed very closely as shown in Figure 5.9(b), thus resulting in a better performance of the RFOC drive.

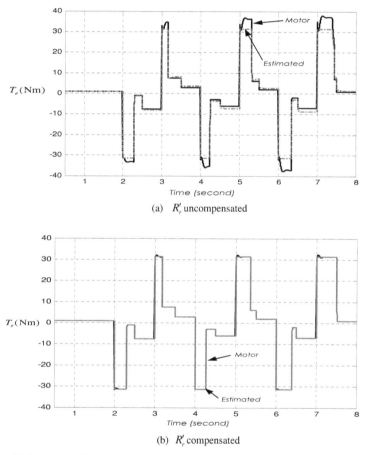

(a) R_r' uncompensated

(b) R_r' compensated

Figure 5.9 Comparison of actual and estimated motor torques for 40% ramp change in R_r; with and without R_r' compensation using ANN (SRIM).

5.4.2 Experimental results with slip-ring induction motor

In order to verify the rotor resistance estimation, a rotor flux oriented induction motor drive was implemented in the laboratory as described in Appendix D. External resistance could be added to the rotor circuit of this slip-ring induction motor.

The external resistance was changed in steps so that the total rotor resistance of the induction motor was increased from 2.33 Ω to 4.68 Ω. This series of measurements were carried out when the motor was running at 1000 rpm and drawing 75% of the full load current. The results obtained for this series of measurements are shown in Table 5.1.

TABLE 5.1 ROTOR RESISTANCE MEASUREMENTS – ADDING EXTRA RESISTANCE

Measured rotor resistance R_r	Estimated rotor resistance $\hat{R}_r$ using the proposed estimator	Percentage error in estimated rotor resistance
2.33 Ω	2.26 Ω	3.00
2.71 Ω	2.59 Ω	4.42
3.82 Ω	3.75 Ω	1.83
4.68 Ω	4.56 Ω	2.56

Immediately after changing the rotor resistance, the estimated rotor resistance was read from the rotor resistance estimator in the controller. After switching off the inverter, this external rotor resistance was measured and added to the rotor resistance of the motor to give the total rotor resistance. This was repeated for three different resistances. The estimated rotor resistance is plotted against the measured total rotor resistance of the induction motor, as shown in Figure 5.10.

Subsequently, a heat run test was conducted for the slip ring motor with 75% full load current at a motor speed of 1000 rpm. The heat run was started at 1.45pm and the stopped at 3.30pm. The estimated rotor resistance was noted at 2.30pm, and the input supply to the inverter was switched off. The rotor resistance was measured immediately. The input supply was switched on again and the heat run test continued

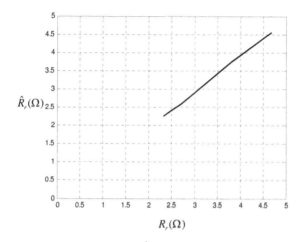

Figure 5.10 Measured (R_r) vs. estimated ($\hat{R}_r$) rotor resistances of the slip-ring induction motor - experimental results.

to take the measurements at 3.00pm. The rotor resistance measurements were repeated as in the previous case and recorded. The final value of rotor resistance was measured at 3.30 pm. Both the estimated rotor resistance using the ANN method and the measured rotor resistance using dc measurements are plotted, as shown in Figure 5.11.

TABLE 5.2 ROTOR RESISTANCE MEASUREMENTS- HEAT RUN

Measured rotor resistance R_r		Estimated rotor resistance using the proposed estimator	Percentage error in estimation
start	2.33 Ω	2.26 Ω	3.00
after 45 minutes	3.10 Ω	2.98 Ω	3.87
after 75 minutes	3.21 Ω	3.08 Ω	4.04
after 105 minutes	3.29 Ω	3.16 Ω	3.95

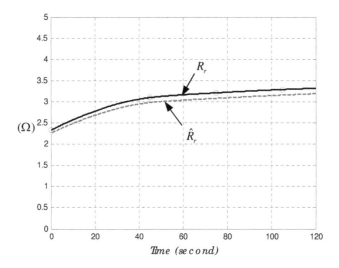

Figure 5.11 Measured and estimated rotor resistances of the slip-ring induction motor during heat run test-experimental results.

5.5 Results with squirrel-cage induction motor

After establishing the validity of the proposed rotor resistance estimation with the slip-ring induction motor, investigations were repeated with a 1.1 kW squirrel-cage induction motor. The parameters of the squirrel-cage induction motor used are given in Table B.1 and the procedure for finding the parameters are described in Appendix-B.2. Further investigations in this book are carried out only for this motor.

5.5.1 Modeling results with squirrel-cage induction motor

To investigate the rotor resistance estimation using ANN, dynamic simulations were performed using the SIMULINK models for the RFOC drive. A schematic diagram showing the RFOC controller, the rotor resistance estimator and the induction motor together with the loading arrangement is shown in Figure 5.12. The stator voltages and currents are measured to estimate the rotor flux linkages using the voltage model as shown in this figure. The inputs to the Rotor Resistance Estimator (RRE)

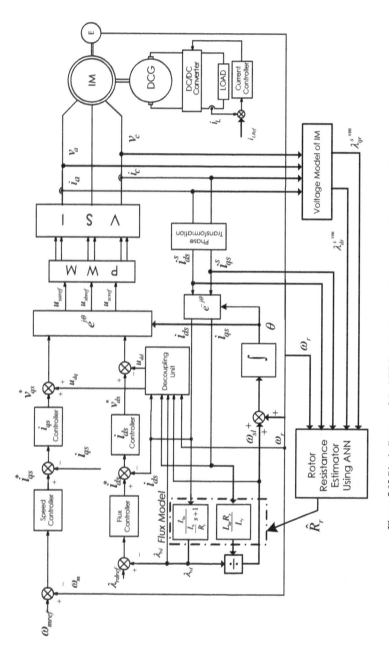

Figure 5.12 Block diagram of the RFOC squirrel-cage induction motor drive with on-line rotor resistance tracking using ANN.

are the stator currents, rotor flux linkages $\lambda_{dr}^{s\ vm}, \lambda_{qr}^{s\ vm}$ and the rotor speed ω_r. The estimated rotor resistance $\hat{R}_r$ will then be used in the RFOC controllers for the flux model.

The performance of the drive system was analyzed by applying an abrupt change in R_r of motor, from 6.03 Ω to 8.5 Ω at 0.8 seconds, whereas the rotor resistance used in the RFOC controller R_r' was kept unaltered. The block - RRE using the ANN in Figure 5.12 was kept disabled in the simulations, to study the effects of change in

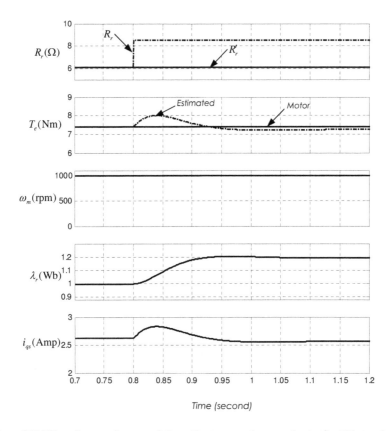

Figure 5.13 Effect of rotor resistance variation without rotor resistance estimator for 40% step change in R_r - modeling results (SCIM).

R_r. The speed of the motor was kept constant at 1000 rpm. The changes in the estimated motor torque T_e, the rotor flux linkage λ_{rd}, and the current i_{qs} were noted. These results are shown in Figure 5.13. The estimated torque was found to drop by nearly 5% and the estimated rotor flux linkage λ_{rd} was found to increase by nearly 21%. The rotor flux increase was only due to the fact that a linear magnetic circuit was assumed without magnetic saturation. The current i_{qs} was also found to follow a similar profile of the motor torque.

Subsequently simulation was repeated after enabling the RRE block, so that the rotor resistance in the controller R_r' was updated with the estimated rotor resistance $\hat{R}_r$. The modeling results obtained for this case, is presented in Figure 5.14. The

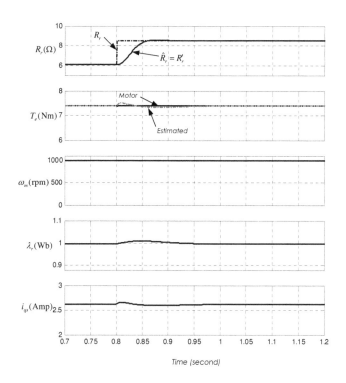

Figure 5.14 Effect of rotor resistance variation with rotor resistance estimator using ANN for 40% step change in R_r - modeling results (SCIM).

estimated rotor resistance $\hat{R}_r$ has converged to the rotor resistance of the motor R_r within 50 milliseconds. The rotor resistance estimator used a sampling interval of 100 microseconds. The rotor flux linkage λ_{rd} was found to remain constant at 1.0 Wb during this transient condition. The current i_{qs} also remained constant as the torque is now perfectly decoupled. Now there is only a small transient disturbance in the rotor flux linkage λ_{rd} and estimated motor torque T_e.

As discussed for the slip-ring induction motor, the rate of temperature rise for the squirrel-cage induction motor is also very slow. The change in rotor resistance due to temperature rise is also very slow. To investigate this situation, a simulation was also carried out introducing a 40% ramp change in the rotor resistance over 8 seconds.

The performance of the drive system was analyzed by changing the rotor resistance by 40%, where the R_r was increased from 6.03 Ω to 8.5 Ω over a period of 8 seconds as shown in the upper trace, without altering R_r' the rotor resistance used in the RFOC controllers. The block - Rotor Resistance Estimator using ANN in Figure 5.12 was kept disabled in the simulations, to study the effects of change in R_r. The speed reference to the drive was kept constant at +1000 rpm for two seconds and then suddenly changed to -200 rpm. Subsequently the speed reference was changed to +400 rpm at 3 seconds, -600 rpm at 4 seconds, 700 rpm at 5 seconds, -800 rpm at 6 seconds and +1000 rpm at 7 seconds. In addition the load torque was kept at 1.0 Nm until 2.5 seconds, and then changed to 7.4 Nm at 2.5 seconds, 3.0 Nm at 3.5 seconds, 6.0 Nm at 4.5 seconds, 2.0 Nm at 5.5 seconds, 7.0 Nm at 6.5 seconds and 1.0 Nm at 7.5 second. The changes in the estimated motor torque T_e, the rotor flux linkage λ_{rd}, and the current i_{qs} were noted. These results are shown in Figure 5.15. The estimated torques were found to be different from the motor torques, as seen in the second trace from the top. The rotor flux linkage λ_{rd} was found to increase as the load torque was increased as shown in this figure, instead of maintaining a constant value of 1.0 Wb. The current i_{qs} followed the same profile as that of the torque.

Subsequently, the simulation was repeated for the same speed references and load torques mentioned in the previous paragraph, after enabling the RRE block, so that the rotor resistance in the controller R'_r was updated with the estimated rotor

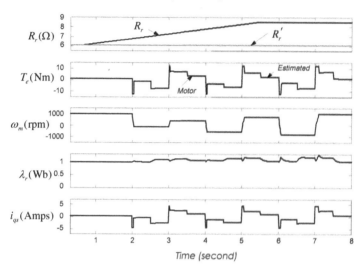

Figure 5.15 Effect of rotor resistance variation without rotor resistance estimator for 40% ramp change in R_r - modeling results (SCIM).

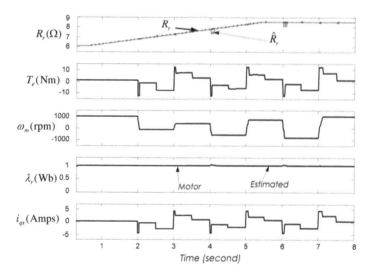

Figure 5.16 Effect of rotor resistance variation with rotor resistance estimator using ANN for 40% ramp change in R_r - modeling results (SCIM).

resistance $\hat{R}_r$. The results obtained for this case, are presented in Figure 5.16. The estimated rotor resistance $\hat{R}_r$ tracked the actual rotor resistance of the motor R_r throughout except there was some small error during the reversal of the motor. The rotor flux linkage λ_{rd} was found to remain constant at 1.0 Wb during this transient condition. The current i_{qs} also remained constant as the torque is now perfectly decoupled. The sampling interval of the rotor resistance estimator was 100 microseconds.

In order to show the difference between the actual and the estimated torques for the uncompensated and compensated cases described by Figures 5.15 and 5.16, they are plotted in a separate enlarged Figure 5.17. In the uncompensated case Figure 5.17(a), there is an error between estimated motor torque and the actual motor torque in the

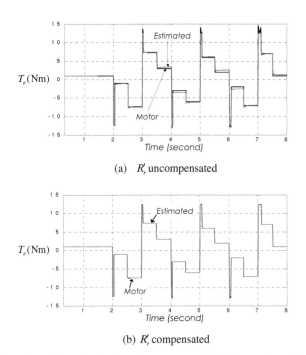

(a) R_r' uncompensated

(b) R_r' compensated

Figure 5.17 Comparison of actual and estimated motor torques for 40% ramp change in R_r; with and without R_r' compensation (SCIM).

94

steady-state. However, when the ANN based rotor resistance was switched on, the estimated and the real motor torques followed very well as shown in Figure 5.17(b), thus resulting in a better performance of the RFOC drive.

5.5.2 Experimental results with squirrel-cage induction motor

In order to examine the capability of tracking the rotor resistance of the induction motor with the proposed estimator, a temperature rise test was conducted, at a motor speed of 1000 rev/min. The results of the R_r estimation obtained from the experiment is shown in Figure 5.18, after logging the data for 60 minutes. Figure 5.19 shows the d–axis rotor flux linkages of the current model (λ_{dr}^{im}), the voltage model (λ_{dr}^{vm}) and the neural model (λ_{dr}^{nm}), taken at the end of heat run. All of the flux linkages are in the stationary reference frame. The flux linkages λ_{dr}^{im} and λ_{dr}^{vm} are updated with a sampling time of 100 μsecond, whereas the flux λ_{dr}^{nm} is updated only at 1000 μsecond. The flux linkage λ_{dr}^{nm} follows the flux linkage λ_{dr}^{vm}, due to the on-line training of the neural network. The coefficients used for training are, $\eta_1 = 0.005$ and $\alpha_1 = 10.0e-6$.

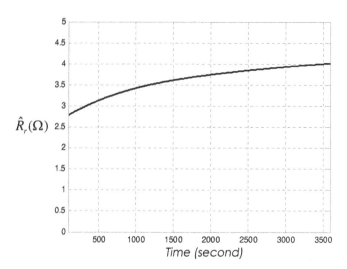

Figure 5.18 Estimated R_r using ANN – experimental results.

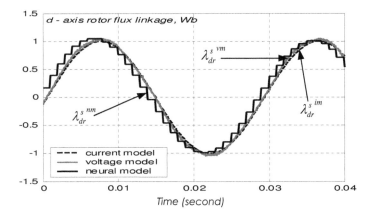

Figure 5.19 Rotor fluxes in R_r estimation using ANN – experimental results.

5.6 Conclusions

This chapter has presented a new rotor resistance identification algorithm using a neural networks principle, for use in indirect vector controlled induction motor drives. It has been validated using a slip-ring induction motor driven from an RFOC controller. The estimator input variables are only stator voltages, stator currents and the motor speed. The simulation and experimental results explained in the previous sections, confirmed the tracking performance of the proposed ANN estimator. Experimental results of R_r variation in the squirrel-cage induction motor is as expected.

CHAPTER 6

ROTOR FLUX ESTIMATION USING PROGRAMMABLE CASCADED LOW-PASS FILTER

6.1 Introduction

The rotor resistance estimation technique using PI / fuzzy estimators described in Sections 4.4 and 4.5 and the technique using artificial neural networks described in Section 5.3 have used the rotor flux linkages estimated with the induction motor voltage model equations. The stator flux can be estimated from the integration of the back-emf by using a single stage integrator. The stator voltages and currents are measured using voltage and current sensors. Operational amplifiers are used to amplify these signals. These sensors have dc offset even when instrumentation amplifiers with very low dc offset and drift is used. The source of offset error is the thermal offset of analog integrators. The dc-offset in the measured stator voltages and stator currents when integrated would eventually lead to very large drifts in the stator flux linkage a runaway problem. The problems when using an integrator for voltage model flux estimation has been investigated by Hinakkanen *et al* [65] and Hu *et al* [66]. Bose *et al* [67] have reported the successful implementation of a stator flux oriented induction motor drive using a programmable cascaded low-pass filter replacing the integrator. Also, Haque *et al* [68] have used the programmable cascaded low-pass filter for compensation of offset error in a Direct Torque Controlled interior permanent synchronous motor drive successfully. Holtz *et al* have proposed a new stator flux estimator obtained by pure integration [42]. This method incorporated a time-varying vector that represents the offset voltages. They achieved higher bandwidth for the stator flux estimation, particularly advantageous when operating the drive at very low speed.

In this book a programmable cascaded low-pass filter approach has been used. Section 6.2 presents a review of voltage model flux estimators, mainly because this was not presented in Chapter 2 because of lack of context at that point. Section 6.3 describes the analysis of the PCLPF. The structure for application of PCLPF for the voltage model flux estimator in a RFOC drive is presented in Section 6.4.

6.2 Brief review of voltage model flux estimation techniques

There has been a substantial amount of research in the development of rotor flux observers for direct field orientation. There was an implementation of a full order rotor flux and stator current closed-loop observer with corrective feedback which can be used to speed up the convergence of the flux estimates and to reduce the sensitivity of the estimates to rotor resistance by Verghese *et al* [69]. The open-loop rotor flux observers and the closed-loop rotor velocity invariant observer were implemented by Lorenz *et al* in [8]. In addition, a new current model flux observer was proposed by Rehman and Xu, which is insensitive to rotor time constant for use in Direct Field Oriented Controller (DFOC) of the induction motor [70]. Another improved closed-loop rotor flux observer using an observer characteristic function was proposed by Sul *et al* for DFOC in [71]. The sensitivity of flux observers to rotor resistance change could not be reduced to zero, and to solve these problems, a rotor flux observer was proposed by Kubota *et al* with a rotor resistance adaptive scheme [72]. The rotor flux observers reported above are compensated for variations in parameters by their feedback corrections.

Recently, a modified integrator with a saturable feedback and an amplitude limiter was proposed by Hu *et al* in [66], to estimate stator flux for drives that require a constant air-gap flux. A modified integrator with adaptive compensation was also proposed in [66] for variable flux operation over a wide speed range. A new, more effective way of calculating the compensation needed together with low-pass filters for rotor flux estimation was presented by Hinkkanen in [65].

Digital implementation of integrators for the estimation of rotor flux of an induction motor from the stator voltages and stator currents, poses problems associated with the offset in the sensor amplifiers and hitherto has not been implemented practically.

Traditional low-pass filters can replace the integrator; however they suffer from poor dynamic response. A Programmable Cascaded Low-Pass Filter (PCLPF) can improve the dynamic performance.

In this chapter, the performance of such an estimator for obtaining rotor flux is further investigated [73]. The rotor flux estimated using this PCLPF has been incorporated into the rotor resistance identification scheme described in Chapters 4 and 5.

6.3 Analysis of flux estimator with a programmable cascaded low-pass filter

For an induction motor, the stator flux linkages in the stator reference frame can be calculated from the measured stator voltages and stator currents using Equations (6.1) and (6.2).

$$\lambda_{ds}^s = \int \left(v_{ds}^s - R_s i_{ds}^s \right) dt \tag{6.1}$$

$$\lambda_{qs}^s = \int \left(v_{qs}^s - R_s i_{qs}^s \right) dt \tag{6.2}$$

Digital implementation of Equations (6.1) and (6.2) can use pure discrete integrators. However, even a small DC offset in the sensors used for the measurement of voltages and currents can produce a very large output after some time as a result of integrating the offset. These integrators can be replaced by low-pass filters. Since the flux estimator has to function over a wide frequency range, a single stage filter has to be designed with a very large time constant, inevitably making it very slow and introducing additional problems. This problem is solved by resolving the single-stage filter into three cascaded low-pass filters with the same time constants and reducing the time constant in each filter by a factor of three, thus the decay time is reduced considerably. These properties of the PCLPF are fully described in [72].

The transfer characteristic of a low-pass filter is given by

$$\frac{Y}{X} = \frac{1}{1 + j\tau\omega} \tag{6.3}$$

where τ is the time constant and ω is the frequency of the signal. The phase lag, Φ, and gain, K, for the filter of Equation (6.3) are:

$$\Phi = tan^{-1}(\tau\omega) \tag{6.4}$$

$$K = \frac{1}{\sqrt{1 + (\tau\omega)^2}} \tag{6.5}$$

As the drive has to operate over a wide speed range, the single stage integrator is resolved into a number of cascaded filters with smaller time constants, so that dc offset decay times are sharply attenuated. A large number of filter sections are desirable, however to minimize the software computation time, a cascade of only three identical filters is selected, in this investigation.

When three low pass filters are cascaded as shown in Figure 6.1, the total phase lag, Φ_T, and total gain, K_T, are expressed as in Equations (6.9) and (6.10) respectively.

$$\Phi_T = tan^{-1}(\tau_1\omega) + tan^{-1}(\tau_2\omega) + tan^{-1}(\tau_3\omega) \tag{6.6}$$

$$K_T = K_1 K_2 K_3 \tag{6.7}$$

When the three filters are identical, $\tau_1 = \tau_2 = \tau_3$,

$$\Phi_T = 3\,tan^{-1}(\tau\omega) \tag{6.8}$$

$$K_T = 3K = \frac{1}{\sqrt{\left[1 + (\tau\omega)^2\right]^3}} \tag{6.9}$$

For the cascaded filter to be equivalent to that of a pure integrator,

$$\Phi_T = 90° \tag{6.10}$$

$$GK_T = \frac{1}{\omega} \tag{6.11}$$

where G is the gain needed for the desired compensation.

Thus,
$$\tau = \frac{tan(90°/3)}{\omega} = f(.)\omega \tag{6.12}$$

$$G = \frac{\sqrt{\left[1+(\tau\omega)^2\right]^3}}{\omega} = g(.)\omega \qquad (6.13)$$

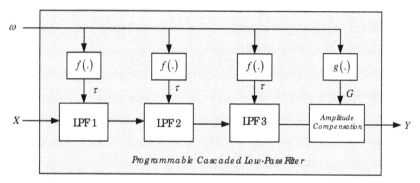

Figure 6.1 Stator flux estimation with programmable cascaded low-pass filters.

Equations (6.12) and (6.13) express the parameters of the filter in terms of the frequency of the input signal ω. Figure 6.2 and Figure 6.3 show the plots of time constant τ and gain G as a function of the stator frequency ω according to the Equations (6.12) and (6.13), respectively.

The cascaded low-pass filter analyzed above can only function as long as the input signal frequency is non-zero. For this case, that is with a dc input signal, τ and G become infinite and the filter fails to perform integration.

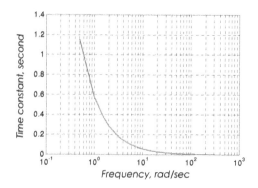

Figure 6.2 Programmable time constant (τ) vs. frequency for the cascaded filter.

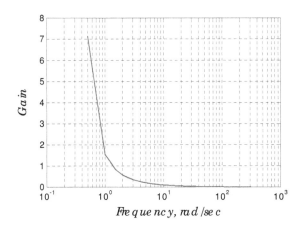

Figure 6.3 Programmable gain (G) vs. frequency for the cascaded filter.

6.4 Application of PCLPF in the RFOC induction motor drive

The instantaneous stator flux linkages $\lambda_{ds}^{s\,vm}, \lambda_{qs}^{s\,vm}$ were estimated using the PCLPF as described in Section 6.3. The instantaneous rotor flux linkages $\lambda_{dr}^{s\,vm}, \lambda_{qr}^{s\,vm}$ are then estimated from the stator flux linkages $\lambda_{ds}^{s\,vm}, \lambda_{qs}^{s\,vm}$ as follows:

$$\lambda_{dr}^{s\,vm} = \frac{L_r}{L_m}\lambda_{ds}^{s\,vm} - \frac{L_s L_r - L_m^2}{L_m}i_{ds}^{s} \tag{6.14}$$

$$\lambda_{qr}^{r\,vm} = \frac{L_r}{L_m}\lambda_{qs}^{s\,vm} - \frac{L_s L_r - L_m^2}{L_m}i_{qs}^{s} \tag{6.15}$$

The back emfs are calculated after subtracting the stator resistance drop from the stator voltages as shown in Figure 6.4. The voltages v_{ds}^{s}, v_{qs}^{s} are the measured PWM stator voltages with necessary hardware filters to get only the sinusoidal voltages and i_{ds}^{s}, i_{qs}^{s} are the measured stator currents.

The instantaneous rotor flux linkages $\lambda_{dr}^{s\,im}, \lambda_{qr}^{s\,im}$ are also computed using the measured stator current and rotor speed, using Equation (6.16), which is referred to as the current model. These two estimated flux linkages $\lambda_{dr}^{s\,vm}$ and $\lambda_{dr}^{s\,im}$ are compared, when the motor is running cold and the parameters are already known.

$$\begin{bmatrix} \dfrac{d\lambda_{dr}^{s\,im}}{dt} \\ \dfrac{d\lambda_{qr}^{s\,im}}{dt} \end{bmatrix} = \begin{bmatrix} -\dfrac{1}{T_r} & -\omega_r \\ \omega_r & -\dfrac{1}{T_r} \end{bmatrix} \begin{bmatrix} \lambda_{dr}^{s\,im} \\ \lambda_{qr}^{s\,im} \end{bmatrix} + \dfrac{L_m}{T_r}\begin{bmatrix} i_{ds}^s \\ i_{qs}^s \end{bmatrix} \qquad (6.16)$$

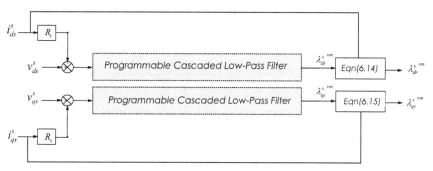

Figure 6.4 Rotor flux estimation with programmable cascaded low-pass filters.

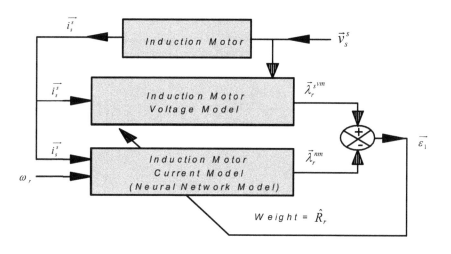

Figure 6.5 Structure of the neural network system for R_r estimation.

This rotor flux vector $\overrightarrow{\lambda_r^{s\,vm}}$ is then used in estimating the rotor resistance using an ANN as already described in Section 5.3. The structure of the neural network system is shown in Figure 6.5.

6.5 Implementation of a three-stage PCLPF and experimental hardware

The programmable cascaded low-pass filter for rotor flux estimation, described in section 6.2 is implemented in software together with a rotor flux oriented induction motor drive as indicated in Figure 6.6. The parameters of the induction motor used are shown in Table B.1. The control system was implemented in software using a DS1102 dSPACE controller board. An IGBT inverter with a switching frequency of 5 kHz was used for the experiment. The real time code was generated with the Real Time Workshop in Matlab, and the dSPACE Real Time Interface Module was used for debugging. The sampling times used for the flux estimation and current controllers were 300 μsecond, and 1200 μsecond for the speed controller.

6.6 Performance evaluation

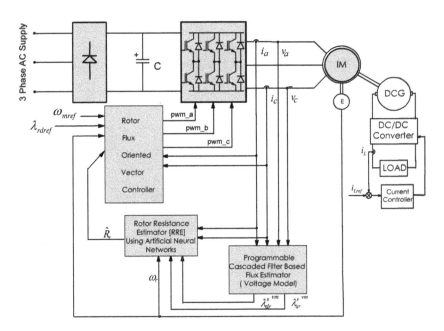

Figure 6.6 Block diagram of the RFOC induction motor drive with ANN based on-line rotor resistance tracking.

104

The cascaded filters are designed and studied with the help of simulations for the IFOC induction motor drive shown in Figure 6.6. The induction motor is controlled with a rotor flux oriented vector controller block. The rotor resistance estimation was carried out using the ANN techniques via the RRE block. The voltage model rotor flux linkages $\lambda_{dr}^{s\,vm}, \lambda_{qr}^{s\,vm}$ are estimated by the PCLPF block shown in this figure, which are then used by the RRE block for rotor resistance estimation. The estimated rotor flux with cascaded filters is compared with the estimated rotor flux using the current model of the induction motor, when the motor is running at an ambient temperature of 20°C.

The performance of the rotor flux estimator using the PCLPF has been studied for four different situations, given below.

A. *Step load-on transients of 7.4Nm* :

When the induction motor was controlled with a speed reference of 1000 rpm, a step load of 7.4 Nm was applied. The simulation results are shown in Figure 6.7(a). The motor torque T_e, the amplitudes of the rotor flux linkage λ_r estimated by PCLPF and the induction motor current model are the two top traces of this figure. The drive has taken nearly 100 milliseconds for the correction in the torque as indicated in the T_e plot. The two bottom traces are the rotor flux linkages λ_{dr}^s along the *d*-axis and λ_{qr}^s along the *q*-axis estimated by the PCLPF and the induction motor current model respectively. There is no error between the rotor flux estimated by the PCLPF and that estimated by the current model.

Figure 6.7(b) shows the corresponding results taken from the experimental set-up. The rotor flux linkages estimated by both the current model and PCLPF have produced the same results as indicated in the bottom two traces. The correction time for the torque loop is again very close to 100 milliseconds.

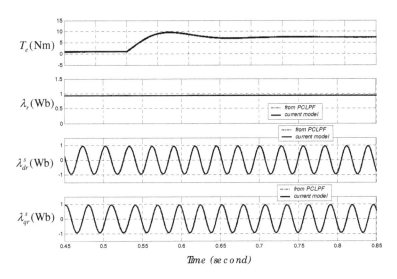

(a) Modeling results

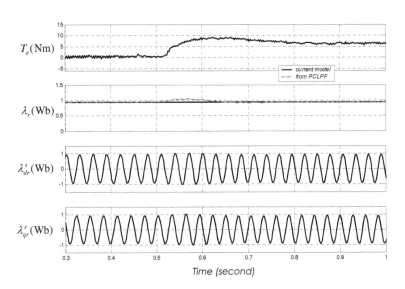

(b) Experimental results

Figure 6.7 Rotor flux estimation results with rated load torque step-on.

B. Step load-off transients of 7.4Nm :

Investigations were also carried out when the load torque of 7.4 Nm was suddenly removed. The modeling results are as indicated in Figure 6.8(a). Here again, there is no error between the rotor flux estimated by the PCLPF and that estimated by the current model. The corresponding results are also taken from the experiment and they are presented in Figure 6.8(b). The rotor flux linkages estimated by the current model and PCLPF are plotted together as the bottom two traces, and they are exactly the same.

C. Step speed transients from 400 to 1000 rpm :

In order to examine the performance of the PCLPF flux estimator for a step change in input frequency, investigations were also carried out for a sudden change in the speed reference. Initially, a step in the speed reference was applied from 400 rpm to 1000 rpm at 0.5 second. The modeling results are presented in Figure 6.9(a). The rotor flux linkages estimated by the PCLPF and the current model have tracked very closely. The experiment was repeated for a corresponding situation and the results are shown in Figure 6.9(b). The rotor fluxes estimated by both the PCLPF and the current model tracked very closely additionally in the experiment.

D. Step speed transients from 1000 to 400 rpm :

The modeling investigation in Part *C* was repeated for a sudden change in the speed reference from 1000 rpm to 400 rpm and the results are shown in the Figures 6.10(a). The difference between rotor flux linkage estimated by the PCLPF and that by the current model during the transients is visible in this case. The results have converged to the same values during the steady-state. Also, experimental results are taken corresponding to this situation and are presented in Figure 6.10(b).

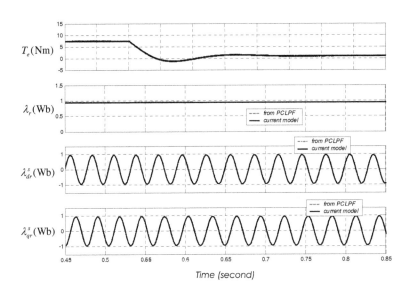

(a) Modeling results

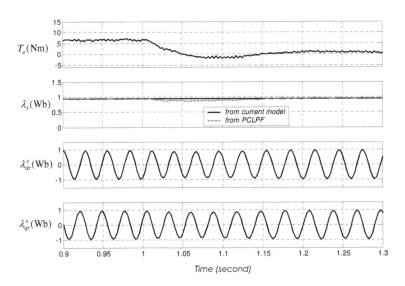

(b) Experimental results

Figure 6.8 Rotor flux estimation results with rated load torque step-off.

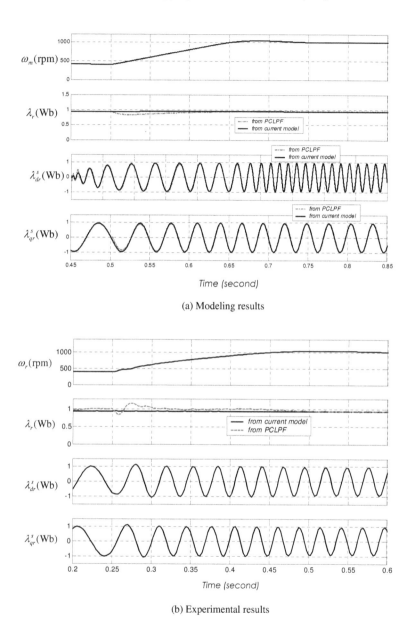

(a) Modeling results

(b) Experimental results

Figure 6.9 Rotor flux estimation results for a step change in speed from 400 rpm to 1000 rpm.

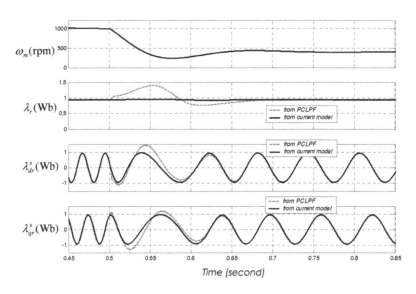

(a) Modeling results

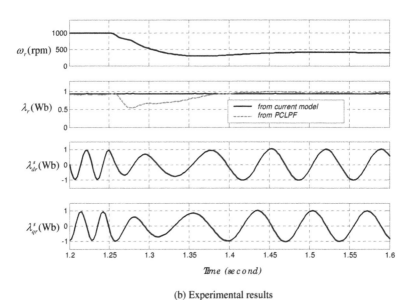

(b) Experimental results

Figure 6.10 Rotor flux estimation results for a step change in speed from 1000 rpm to 400 rpm.

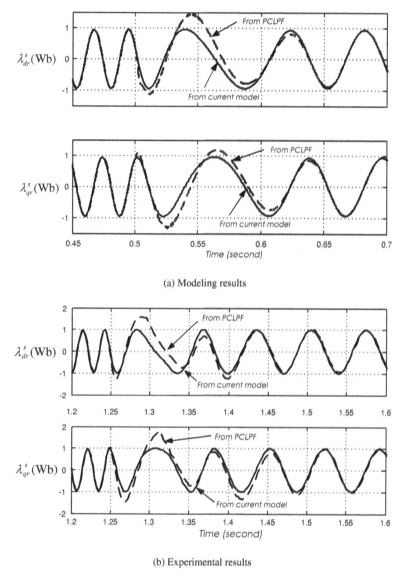

(a) Modeling results

(b) Experimental results

Figure 6.11 Rotor flux estimation results during transients from 1000 rpm to 400 rpm.

In order to understand the flux estimation during the above transients the results of Figure 6.10(a) is again plotted as Figure 6.11(a) only for the time interval 0.45 second to 0.7 second. The rotor flux linkages estimated by the PCLPF and by the current model are shown to converge within one input cycle. Also the transient results of Figure 6.10(b) are plotted again as Figure 6.11(b), for the time interval 1.2 second to 1.6 second. The fluxes estimated by the PCLPF flux estimator have converged to the current model fluxes within nearly one input cycle.

6.7   Conclusions

This chapter has presented the suitability of a programmable cascaded low-pass filter for rotor flux estimation using measured stator voltages and stator currents, for the purpose of rotor resistance identification. The rotor flux has been shown to be estimated sufficiently accurately for use in an IFOC drive. The filter estimates the rotor flux for both steady state and transient conditions of an induction motor drive successfully. The flux estimation results of the PCLPF method were in total agreement with the rotor flux estimated by the current model during both steady state and transient conditions. The proposed PCLPF method does not require any arbitrary selection of gains or tuning as reported in [65] and [66], and consequently the PCLPF offers some simplicity with respect to methods available in the literature.

CHAPTER 7

STATOR RESISTANCE ESTIMATION WITH A PI ESTIMATOR AND FUZZY LOGIC

7.1   Introduction

The R_r estimation algorithm reported in chapter 5 requires the knowledge of stator resistance (R_s) which may vary up to 50% during operation of the drive. Even though R_s is not included in the controller for the RFOC drive, its variation affects the estimate of R_r. It has been observed in Marino's paper [38] and also confirmed experimentally in Section 7.2.2 that the error in R_s, leads to errors in R_r estimation. This aspect has not been taken into account individually in the closed-loop estimates for R_r. The rotor resistance estimator described in this section, has used the fluxes $\lambda_{dr}^{s\ vm}$, $\lambda_{qr}^{s\ vm}$ derived from the voltage model of the induction motor. This flux estimation is dependent on the stator resistance R_s of the induction motor as shown by Equation (5.1). Modeling results in Section 7.2.1.2 clearly shows that maximum possible variation in R_s introduces an error in the estimate of R_r. Furthermore, the error in λ_r^s estimation due to variation in R_s is known to introduce significant error in speed estimation for a speed sensorless drive, which will be described in Chapter 9. These provide the incentive for R_s estimation. In order to minimize the error in rotor resistance estimation, resulting from the stator resistance variation, an on-line stator resistance estimator is integrated which is implemented with a simple PI controller or a fuzzy logic system in this chapter.

7.2   Stator resistance estimation using a PI compensator

The rotor flux estimations using the voltage model equation (5.1), and current model equation (5.2), can also be written as:

$$\sigma L_s \frac{di_{ds}^s}{dt} = \frac{L_m}{L_r T_r} \lambda_{dr}^{s\ im} + \frac{L_m}{L_r} \omega_r \lambda_{qr}^{s\ im} - \frac{L_m^2}{L_r T_r} i_{ds}^s + v_{ds}^s - R_s i_{ds}^s \tag{7.1}$$

$$\sigma L_s \frac{di_{qs}^s}{dt} = \frac{L_m}{L_r T_r} \lambda_{qr}^{s\ im} - \frac{L_m}{L_r} \omega_r \lambda_{dr}^{s\ im} - \frac{L_m^2}{L_r T_r} i_{qs}^s + v_{qs}^s - R_s i_{qs}^s \tag{7.2}$$

Using the discrete form of equation (7.1), the estimated d–axis stator current can be represented as:

$$i_{ds}^{s*}(k) = W_4 i_{ds}^{s*}(k-1) + W_5 \lambda_{dr}^{s\ im}(k-1) + W_6 \omega_r \lambda_{qr}^{s\ im}(k-1) + W_7 v_{ds}^s(k-1) \tag{7.3}$$

where the coefficients are,

$$W_4 = 1 - \frac{T_s}{\sigma L_s} \frac{L_m^2}{L_r T_r} - \frac{T_s}{\sigma L_s} R_s ; \quad W_5 = \frac{T_s}{\sigma L_s} \frac{L_m}{L_r T_r}$$

$$W_6 = \frac{T_s}{\sigma L_s} \frac{L_m}{L_r} ; \quad W_7 = \frac{T_s}{\sigma L_s}$$

Similarly, using the discrete form of (7.2),

$$i_{qs}^{s*}(k) = W_4 i_{qs}^{s*}(k-1) + W_5 \lambda_{qr}^{s\ im}(k-1) - W_6 \omega_r \lambda_{dr}^{s\ im}(k-1) + W_7 V_{qs}(k-1) \tag{7.4}$$

The amplitude of the stator current is represented as:

$$I_s^*(k) = \sqrt{i_{ds}^{s*}(k)^2 + i_{qs}^{s*}(k)^2} \tag{7.5}$$

Similarly, the amplitude of the measured stator current is calculated as:

$$I_s(k) = \sqrt{i_{ds}^s(k)^2 + i_{qs}^s(k)^2} \tag{7.6}$$

Equation (7.5) estimates the stator current for the condition stator resistance is equal to R_s. Subsequently, if the stator resistance changes by ΔR_s, that is, the new stator resistance becomes $R_s + \Delta R_s$. The corresponding new stator current estimate will be given by equation (7.7).

$$I_s^*(k) + \Delta I_s^*(k) = \sqrt{\left(i_{ds}^{s*}(k) + \Delta i_{ds}^*(k)\right)^2 + \left(i_{qs}^{s*} + \Delta i_{qs}^{s*}\right)^2} \tag{7.7}$$

The error ΔR_s introduced in R_s has developed an error in the stator current estimate ΔI_s^*.

This error can be taken through a PI estimator to recover the change in resistance ΔR_s as given by (7.7), where K_p is the proportional gain and K_i is the integral gain of the PI estimator.

$$\Delta R_s = \left(K_p + \frac{K_i}{s} \right) \Delta I_s^* \tag{7.7}$$

To examine the effect of stator resistance variation in the amplitude of stator current, modeling studies were carried out with a trapezoidal variation in stator resistance. The induction motor is controlled with a rotor flux oriented controller, loaded with rated torque of 7.4 Nm. The initial value of stator resistance was kept at 6.03 Ω. Then the stator resistance was increased to 8.0 Ω and subsequently dropped to the nominal value of 6.03 Ω. The stator current profile for this cycle is plotted and is shown in Fig. 7.1. The relationship between stator current and stator resistance is non-linear. The hysteresis observed in the stator current is due to the time delay of the low-pass filter which is used to remove the ripples in the current.

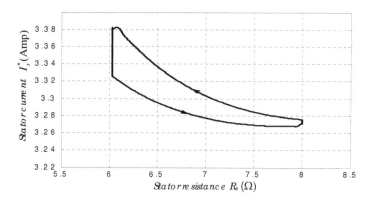

Figure 7.1 Relationship between variations of amplitude of stator current with R_s.

In the proposed PI stator resistance estimator, shown in Figure 7.2, the error between the estimated stator current I_s^* and the current I_s measured from the motor is used to determine the incremental value of stator resistance through a PI controller and limiter [82]. The amplitude of the stator current I_s^* is estimated using the Equations

(7.3), (7.4) and (7.5). The inputs to this estimator block are the measured stator voltages v_{ds}^s, v_{qs}^s along the d and q axes in the stationary reference frame, the rotor speed ω_r and the current model rotor flux linkages $\lambda_{dr}^{s\,im}, \lambda_{qr}^{s\,im}$. The measured stator current amplitude I_s is the subtracted from I_s^*. This current error is taken through a low-pass filter, which has a very low cutoff frequency in order to remove high frequency components contained in the measured stator current. This filter time constant does not generate any adverse effect on the stator resistance adaptation if the filter time constant is chosen to be smaller than that of the adaptation time constant. The incremental stator resistance, $\Delta \hat{R}_s$ is then continuously added to the previously estimated stator resistance, $\hat{R}_{s0}$. The final estimated value, $\hat{R}_s$, is obtained as the output of another low-pass filter and limiter. The low-pass filter is necessary for a smooth variation of the estimated resistance value. This final value $\hat{R}_s$ is the updated stator resistance and is used as the stator resistance R_s in the equation to calculate W_4.

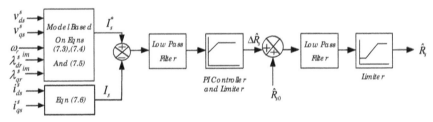

Figure 7.2 Block diagram of R_s identification using PI estimator.

7.2.1 Modeling results of PI stator resistance estimator

The block diagram of a rotor flux oriented induction motor drive, together with both stator and rotor resistance identifications, is shown in Figure 7.3. The induction motor is controlled with a Rotor Flux Oriented Vector Controller as shown in this figure. The voltage model fluxes are estimated from the measured stator voltages and currents using the PCLPF. The flux estimation using the PCLPF has already been described in Chapter 6. The stator voltages are PWM voltages and are filtered with hardware filters on the voltage sensor boards inside the IGBT inverter and only the sinusoidal voltages are taken to the PCLPF flux estimators. The rotor resistance is estimated using artificial neural networks by the *Rotor Resistance Estimator* (RRE) block. The operation of this block has already been described in Chapter 5. The stator resistance estimation described in Section 7.2 is implemented by the *Stator Resistance Estimator* (SRE) block shown in Figure 7.3. The implementation of rotor resistance identification with an ANN supplemented with stator resistance estimation using the PI compensator of Figure 7.2, has been verified by modeling studies with SIMULINK software.

In order to investigate the performance of the drive for parameter variations in rotor resistance R_r, a series of simulations were conducted by introducing error between the actual value R_r and the value used in the controller R'_r. Similarly, another series

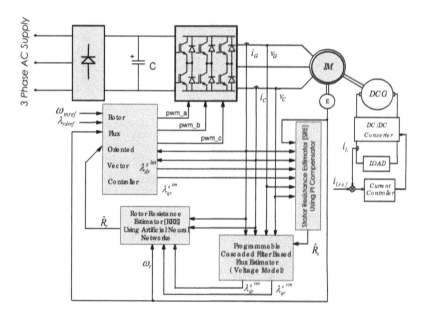

Figure 7.3 Block diagram of the RFOC induction motor drive with on-line ANN rotor resistance estimation supplemented with a PI stator resistance estimator.

of simulations were conducted by introducing error between the actual stator resistance R_s and the one used in the estimator R'_s. All these investigations were conducted for the drive running at 1000 rev/minute and a constant load torque of 7.4 Nm. The parameters of the motor used for modeling studies are in Table B.1.

7.2.1.1   With RRE and SRE off

Initially R_r and R_s were increased abruptly by 40% at 1.5 second, keeping both the Rotor Resistance Estimator (RRE) and Stator Resistance Estimator (SRE) blocks off in Figure 7.3. The steady-state values of the torque, rotor flux linkage and the amplitude of the stator current vector are shown in Figure 7.4. The rotor flux linkage in the motor increases by 21% compared to its estimated value, when the error in rotor resistance is introduced, as shown in Figure 7.4(iv). The estimated torque is 4% lower than the actual motor torque, as shown in Figure 7.4(ii). Also there is a 3.25% drop in the amplitude of the stator current vector starting at 1.5 second, when the error is introduced, as shown in Figure 7.4(vi).

7.2.1.2   With RRE on and SRE off

Later, simulations were repeated after switching on only the ANN Rotor Resistance Estimation block with the SRE block still off, for the same changes in resistances which were introduced in Figure 7.4. The estimated $\hat{R}_r$ in this case is higher than the R_r by 1.7% as shown in Figure 7.5(i). The estimated torque is 1.0% higher than the real motor torque, as shown in Figure 7.5(ii). However, the estimated rotor flux linkage is 1.27% lower than the actual rotor flux linkage as indicated in Figure 7.5(iv). The stator current amplitude increases only by 0.4% in this case, as shown in Figure 7.5(vi).

7.2.1.3   Both RRE and SRE on

Finally, simulations were carried out with both the RRE and SRE blocks switched on for the same changes in R_r and R_s. The results of torque, rotor flux linkage and stator current amplitude are shown for both of the cases, in Figure 7.5. The errors reported in the previous paragraph, between estimated and real quantities of torque rotor flux linkage and stator current amplitude have largely disappeared in this case. The estimated rotor resistance has tracked the real rotor resistance of the motor very well, as the estimation error now drops to 0.4% as shown in Figure 7.5(i). The error in the estimated stator resistance $\hat{R}_s$ is now very close to zero.

Figures 7.4 and 7.5 describe the possible steady-state errors encountered in a situation where a step change in resistance is applied, only for the purpose of investigation and

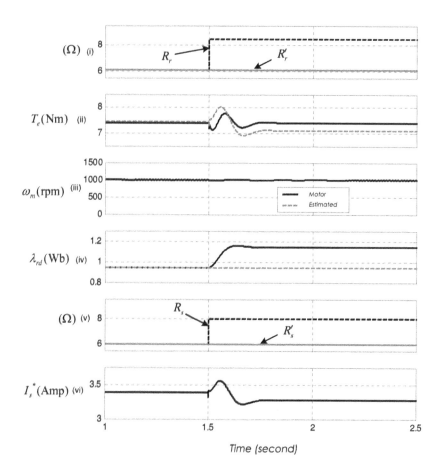

Figure 7.4 Performance of the drive without rotor and stator resistance estimations for 40 % step change in R_r and R_s, R'_r and R'_s uncompensated – modeling results.

verification of the technique. However, the practical variation in resistances is very slow. A corresponding modeling investigation is also carried out, and the results are indicated in Figure 7.6. The simulations are done in three steps. At first the drive

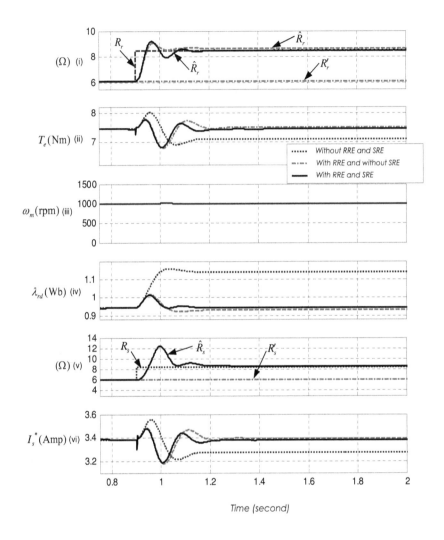

Figure 7.5 Performance of the drive with R_r estimation using ANN and R_s estimation using PI compensator for 40 % step change in R_r and R_s, R_r' and R_s' compensated – modeling results.

system is analyzed after introducing error between R_r and $\hat{R}_r$ and R_s and $\hat{R}_s$ keeping both RRE and SRE turned off. Repeated simulations were also carried out, with RRE

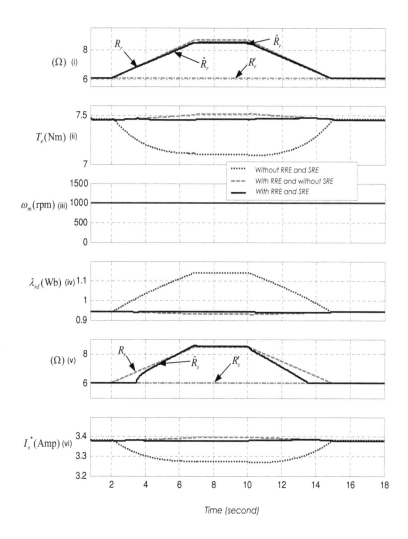

Figure 7.6 Performance of rotor resistance estimation using ANN and stator resistance estimation using PI estimator for 40% ramp change in R_r and R_s, both R'_r and R'_s compensated – modeling results.

on and SRE off. The $\hat{R}_r$ estimated in this case is higher than the R_r by 1.7% as shown in Figure 7.6(i). The estimated torque is 0.9% higher than the real motor torque, as shown in Figure 7.6(ii). However, the estimated rotor flux linkage is 2.1% lower than the actual rotor flux linkage as indicated in Figure 7.6(iv). The stator current amplitude increases only by 0.4% in this case, as shown in Figure 7.6(vi).

Finally, both rotor and stator resistance estimators are investigated with both the RRE and SRE switched on. The estimated rotor resistance has tracked the real rotor resistance of the motor very well, as the error now drops to 0.1% as in Figure 7.6(i). The steady state error of the estimated stator resistance with respect to the real stator resistance is zero, as shown in Figure 7.6(v).

7.2.2 Experimental results of PI stator resistance estimator

In order to verify the proposed stator and rotor resistance estimation algorithms, a rotor flux oriented induction motor drive was implemented in the laboratory as shown in Figure 7.7. The experimental set-up was described in Section 5.5.

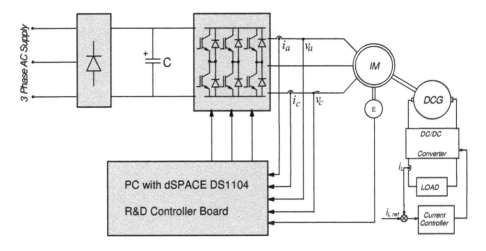

Figure 7.7 Experimental set-up for the resistance identification in induction motor drive.

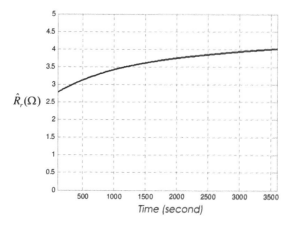

Figure 7.8 Estimated $\hat{R}_r$ using ANN – experimental results.

In order to examine the capability of tracking the rotor resistance of the induction motor with the proposed estimator, a temperature rise test was conducted, at a motor

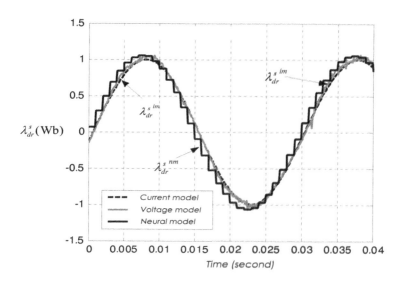

Figure 7.9 Rotor fluxes in R_r estimation; RRE using ANN – experimental results.

speed of 1000 rev/min. The results of estimated $\hat{R}_r$ from the experiment is shown in

Figure 7.8, after logging the data for 60 minutes. Figure 7.9 shows the d–axis rotor flux linkages of the current model ($\lambda_{dr}^{s\ im}$), the voltage model ($\lambda_{dr}^{s\ vm}$) and the neural model ($\lambda_{dr}^{s\ nm}$), taken at the end of heat run. All of the flux linkages are in the stationary reference frame. The flux linkages $\lambda_{dr}^{s\ im}$ and $\lambda_{dr}^{s\ vm}$ are updated with a sampling time of 100 µsecond, whereas the flux $\lambda_{dr}^{s\ nm}$ is updated only at 1000 µsecond. The flux linkage $\lambda_{dr}^{s\ nm}$ follows the flux linkage $\lambda_{dr}^{s\ vm}$, due to the on-line training of the neural network. The coefficients used for training are, η_1 = 0.005 and α_1 =10.0e-6.

In order to test the stator resistance estimation by experiment, an additional 3.4 Ω per phase was added in series with the induction motor stator by opening a circuit breaker shown in Figure E.3, when the motor was running at 1000 rev/min with a load torque of 7.4 Nm. The estimated stator resistance together with the actual stator resistance is shown in Figure 7.10. The estimated stator resistance converges to 9.4 Ω within 150 milliseconds. Figure 7.11 shows both the measured d-axis stator current and the one computed by the PI estimator.

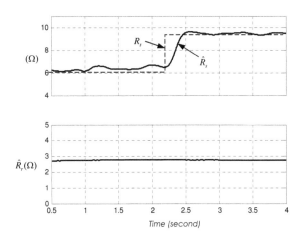

Figure 7.10 Estimated $\hat{R}_s$ and $\hat{R}_r$; RRE using ANN and SRE using PI estimator – experimental results.

124

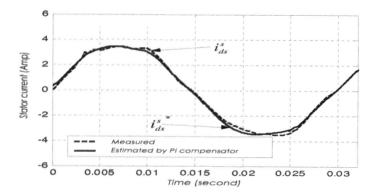

Figure 7.11 Stator currents in R_s estimation; SRE by PI estimator – experimental results.

7.2.3 Analysis of Results

The modeling results as described in Figure 7.5 indicates that the proposed rotor and stator resistance estimators can converge in a short time, as low as 200 milliseconds corresponding to a 40% step change for both stator and rotor resistances

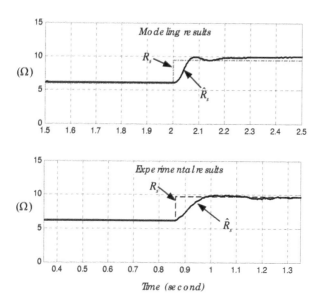

Figure 7.12 Comparison of stator resistance estimations simulation vs. experiment- SRE by PI estimator.

simultaneously. In order to compare the stator resistance estimation for simulation and experiment, simulation is repeated with SRE and RRE blocks switched on in Figure 7.3. Then a step change in R_s is applied without any change in R_r and the results are recorded as the upper trace in Figure 7.12. The bottom trace in Figure 7.12 is the same as the top trace of Figure 7.10. The estimation time as modeled is in very close agreement with that obtained from experiment.

7.3   Stator resistance estimation using fuzzy estimator

The non-linear relationship between stator current and stator resistance discussed in Section 7.1 could be easily mapped using a fuzzy logic system. A fuzzy logic estimator was proposed, which was designed to estimate the change in stator resistance while the drive is in operation, which is shown in Figure 7.13. As described in Section 7.2, the amplitude of the stator current I_s^* is estimated using Equations (7.3), (7.4) and (7.5). The inputs to this estimator block in Figure 7.13 are the measured stator voltages v_{ds}^s, v_{qs}^s along the d and q axes in the stationary reference frame, the rotor speed ω_r and the current model rotor flux linkages $\lambda_{dr}^{s\,im}, \lambda_{qr}^{s\,im}$. The measured stator current amplitude I_s is the subtracted from I_s^*. The error between the estimated stator current $I_s^*(k)$ and the measured stator current $I_s(k)$ is used to determine the incremental value of stator resistance $(\Delta \hat{R}_s)$ through a fuzzy logic estimator [74]. The inputs to the estimator are the current error $e(k)$ and change in current error $\Delta e(k)$.

$$e(k) = \Delta I_s(k) = I_s^*(k) - I_s(k) \qquad (7.8)$$

$$\Delta e(k) = e(k) - e(k-1) \qquad (7.9)$$

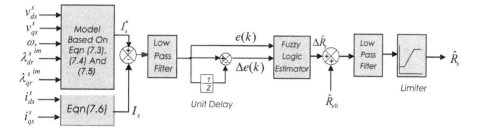

Figure 7.13 Block diagram of R_s identification using fuzzy estimator.

The fuzzification stage input variables for the resistance estimator are current error $e(k)$, change in current error $\Delta e(k)$ and the output variable is the change in stator resistance ΔR_s. These variables are divided into seven fuzzy segments namely where NL, NM, NS, Z, PS, PM and PL which corresponds to Negative Large, Negative Medium, Negative Small, Zero, Positive Small, Positive Medium and Positive Large respectively. The crisp input variables are converted into fuzzy variables using triangular membership functions as shown in Figure 7.15 and Figure 7.16. The range of control, which is the universe of discourse is -0.3 to 0.3 Amps for current error and -0.25 to 0.25 for change in current error respectively. The rule base of the fuzzy logic estimator is shown in Figure 7.14, which has 49 rules.

The interface method used is basic and is developed from the minimum operation rule as the fuzzy implementation function. The firing strength of ith rule is given by

$$\alpha_i = \min\left\{\mu_{e_i}(e), \mu_{\Delta e_i}(\Delta e)\right\} \tag{7.10}$$

where $\mu e_i(e)$ is the membership grade of the e_i segment of current error e and $\mu_{\Delta ei}(\Delta e)$ is the membership grade of the Δe_i segment of change in the current error Δe. The membership function of the resultant aggregation is by the maximum method. The maximum method takes the maximum of their Degree of Support (DoS) values.

		Current error						
		NL	NM	NS	Z	PS	PM	PL
Change in current error	NL	NL	NL	NL	NL	NM	NS	Z
	NM	NL	NL	NL	NM	NS	Z	PS
	NS	NL	NL	NM	NS	Z	PS	PM
	Z	NL	NM	NS	Z	PS	PM	PL
	PS	NM	NS	Z	PS	PM	PL	PL
	PM	NS	Z	PS	PM	PL	PL	PL
	PL	Z	PS	PM	PL	PL	PL	PL

Figure 7.14 Rule base for the fuzzy stator resistance estimator.

In the defuzzification stage, a crisp value for the output variable, change in resistance (ΔR_s) is obtained by using the Mean of Maximum (MoM) operator. The membership functions for the defuzzification stage are shown in Figure 7.17. Figure 7.18 shows the control surface of the fuzzy stator resistance estimator, where the horizontal axes are current error and change in current error and vertical axis is change in stator resistance.

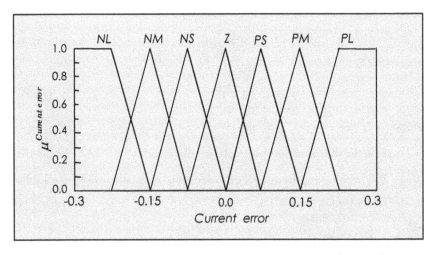

Figure 7.15 Membership distribution of current error for the fuzzy stator resistance estimator.

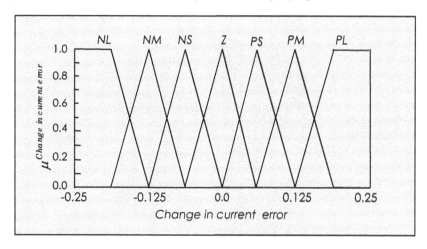

Figure 7.16 Membership distribution of Δ current error for the fuzzy stator resistance estimator.

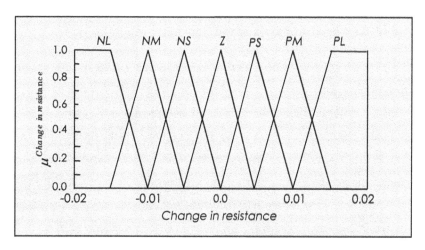

Figure 7.17 Membership distribution for ΔR_s for the fuzzy stator resistance estimator.

The incremental stator resistance $\Delta\hat{R}_s$ is continuously added to the previously estimated stator resistance $\hat{R}_{s0}$. The final estimated value $\hat{R}_s$ is obtained as the output of another low-pass filter and limiter.

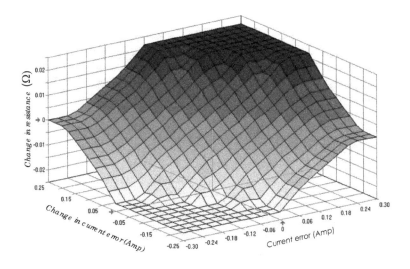

Figure 7.18 Control surface of fuzzy stator resistance estimator.

7.3.1 Modeling results of fuzzy stator resistance estimator

The block diagram of a rotor flux oriented induction motor drive, together with both stator and rotor resistance observers, is shown in Figure 7.19. The induction motor is controlled with a Rotor Flux Oriented Vector Controller as shown in this figure. The voltage model fluxes are estimated from the measured stator voltages and currents using the PCLPF. The flux estimation using the PCLPF is described in Chapter 6. The stator voltages are PWM voltages and are filtered with hardware filters on the voltage sensor boards inside the IGBT inverter and only the sinusoidal voltages are taken to the PCLPF flux estimators. The rotor resistance is estimated using artificial neural networks by the Rotor Resistance Estimator (RRE) block. The operation of this block is already described in Chapter 5. The stator resistance estimation described in Section 7.3 is implemented by the Stator Resistance Estimator (SRE) block shown in Figure 7.19. The implementation of rotor resistance identification with an ANN supplemented with stator resistance estimation using the fuzzy logic estimator of Figure 7.13, has been verified by modeling studies with SIMULINK together with *Inform fuzzy*TECH software [61].

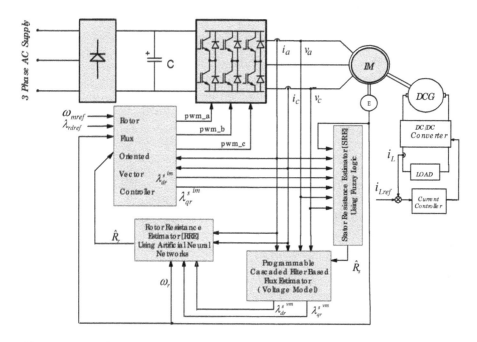

Figure 7.19 Block diagram of RFOC induction motor drive with on-line ANN rotor resistance estimation supplemented with a fuzzy stator resistance estimator.

In order to investigate the performance of the drive for parameter variations in rotor resistance R_r, a series of simulations were conducted by introducing error between the actual value R_r and the value used in the controller R'_r. Similarly, another series of simulations were conducted by introducing error between the actual stator resistance R_s and the one used in the controller R'_s. All of these investigations were conducted for the drive machine running at 1000 rev/minute and with a constant load torque of 7.4 Nm.

7.3.1.1   With RRE and SRE off

These results are the same as the one discussed in Section 7.1.1.1 since there are no resistance estimators in this case.

7.3.1.2   With RRE on and SRE off

Simulations were repeated after switching on only the ANN Rotor Resistance Estimation block with the SRE block still off, for the same changes in resistances which were introduced in Figure 7.4. The estimated $\hat{R}_r$ in this case is higher than the R_r by 1.7% as shown in Figure 7.20(i). The estimated torque is 1.0% higher than the real motor torque, as shown in Figure 7.20(ii).

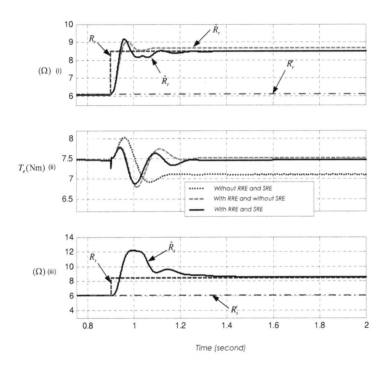

Figure 7.20 Performance of the drive with R_r estimation using ANN and R_s estimation using fuzzy estimator for 40 % step change in R_r and R_s, R'_r and R'_s compensated – modeling results.

7.3.1.3   With RRE and SRE on

Finally, simulations were carried out with both the RRE and SRE blocks switched on for the same changes in R_r and R_s. The result of estimated torque is shown for both of the cases, in Figure 7.6(a). The errors reported in the previous paragraph, between estimated and real quantities of torque have largely disappeared in this case. The estimated rotor resistance has tracked the real rotor resistance of the motor very well, as the estimation error now drops to 0.4% as shown in Figure 7.20(i). The error in the estimated stator resistance $\hat{R}_s$ is now very close to zero, as indicated in Figure 7.20(iii).

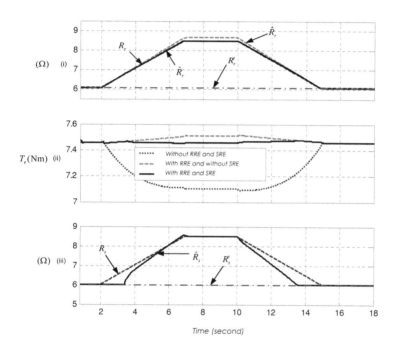

Figure 7.21 Performance of the drive with R_r estimation using ANN and R_s estimation using fuzzy estimator for 40 % ramp change in R_r and R_s, R'_r and R'_s compensated – modeling results.

The results shown by Figure 7.20 described the possible steady-state errors encountered in a situation where a step change in resistance is applied, only for the purpose of investigation and the verification of the technique. However, the practical variation in resistances is very slow. A corresponding modeling investigation is also carried out, and the results are indicated in Figure 7.21. The simulations have been carried out in three steps. Initially R_r and R_s were increased gradually by 40% at 2 second, keeping both the Rotor Resistance Estimator (RRE) and Stator Resistance Estimator (SRE) blocks off in Figure 7.19. The R_s and R_r were allowed to remain steady from 7 seconds to 10 seconds and then reduced gradually back to its original value. Repeated simulations were also carried out after switching on only the RRE on with the SRE block still off. The estimated $\hat{R}_r$ in this case is higher than the R_r by 1.7% as shown in Figure 7.21(i). The estimated torque is 0.9% higher than the real motor torque, as shown in Figure 7.21(ii).

Finally, both rotor and stator resistance estimators were investigated with both RRE and SRE switched on for the same changes in R_r and R_s. The estimated rotor resistance has tracked the real rotor resistance of the motor very well, as the error now drops to 0.1% as in Figure 7.21(i). The steady state error of the estimated stator resistance with respect to the real stator resistance is zero, as shown Figure 7.21(iii).

7.3.2 Experimental results of fuzzy stator resistance estimator

In order to verify the proposed stator and rotor resistance estimation algorithms, a rotor flux oriented induction motor drive shown in Figure 7.19 was implemented in the laboratory. The details of the experimental set-up are described in Appendix-D.

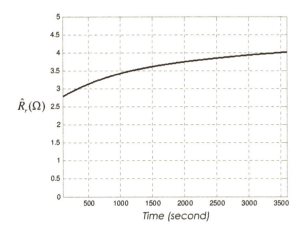

Figure 7.22 Estimation of R_r using ANN – experimental results.

In order to examine the capability of tracking the rotor resistance of the induction motor with the proposed estimator, a temperature rise test was conducted, at a motor speed of 1000 rev/min. The results of estimated $\hat{R}_r$ from the experiment is shown in Figure 7.22, after logging the data for 60 minutes. Figure 7.23 shows the d–axis rotor flux linkages of the current model$\left(\lambda_{dr}^{s\ im}\right)$, the voltage model $\left(\lambda_{dr}^{s\ vm}\right)$ and the neural model$\left(\lambda_{dr}^{s\ nm}\right)$, taken at the end of heat run. All of the flux linkages are in the stationary reference frame. The flux linkages $\lambda_{dr}^{s\ im}$ and $\lambda_{dr}^{s\ vm}$ are updated with a sampling time of 100 μsecond, whereas the flux $\lambda_{dr}^{s\ nm}$ is updated only at 1000 μsecond. The flux linkage $\lambda_{dr}^{s\ nm}$ follows the flux linkage $\lambda_{dr}^{s\ vm}$, due to the on-line training of the neural network. The coefficients used for training are, $\eta 1 = 0.005$ and $\alpha 1 = 10.0e\text{-}6$.

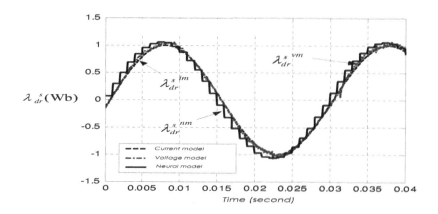

Figure 7.23 Rotor fluxes in R_r estimation; RRE using ANN – experimental results.

To examine the effect of stator resistance in the rotor resistance estimation, an additional 3.4 Ω per phase was added in series with the induction motor stator, with the drive operated with RFOC, and the *RRE* on and *SRE* off. The estimated rotor resistance $\hat{R}_r$ was found to increase from 2.78 Ω to 2.83 Ω, as indicated in Figure 7.24.

To verify the stator resistance estimation together with the rotor resistance estimation, an additional 3.4 Ω per phase was added in series with the induction motor stator, with the motor running at 1000 rev/min and with a load torque of 7.4 Nm. The estimated stator resistance together with the actual stator resistance is shown in Figure 7.26. The estimated stator resistance converges to 9.4 Ω within 150 milliseconds. Figure 7.25 shows both the measured d-axis stator current and the one computed by the fuzzy estimator.

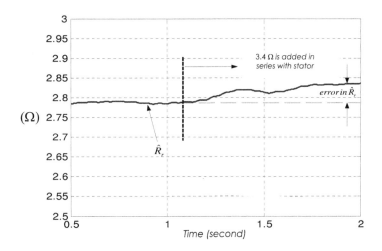

Figure 7.24 Estimated $\hat{R}_r$ with SRE disabled for step change in R_s – experimental results.

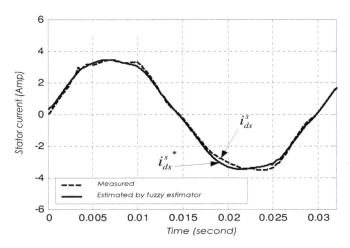

Figure 7.25 Stator currents in R_s estimation with SRE using fuzzy estimator – experimental results.

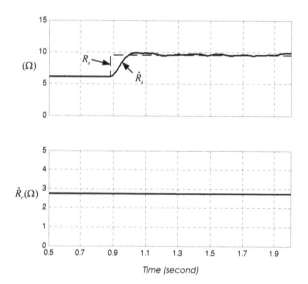

Figure 7.26 Estimated $\hat{R}_r$ and $\hat{R}_s$ with RRE using ANN estimator and SRE using fuzzy estimator – experimental results.

7.3.3 Analysis of Results

The modeling results as described in Figure 7.20 indicate that the proposed rotor and stator resistance estimators can converge in a short time, as low as 200 millisecond corresponding to a 40% step change for both stator and rotor resistances simultaneously. In order to compare the stator resistance estimation for simulation and experiment, simulation is repeated with *SRE* and *RRE* blocks switched on in Figure 7.19. Then a step change in R_s is applied without any change in R_r and the results are recorded as the upper trace in Figure 7.27. The bottom trace in this figure is the same as the top trace of Figure 7.26. The estimation time in the modeling is in very close agreement with that obtained from the experiment.

7.4   Conclusions

This chapter has presented a new on-line estimation technique for the rotor resistance R_r in the presence of R_s variations for the induction motor drive. As a result of the

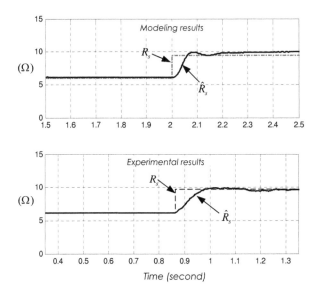

Figure 7.27 Comparison of stator resistance estimations - simulation vs. experiment; fuzzy estimation.

stator resistance estimation which is embedded separately, the R_r estimation was found to be totally insensitive to R_s variations.

Investigations carried out in this chapter have clearly shown that an ANN can be used to estimate R_r, and its dependence on stator resistance can be easily reduced significantly using a PI estimator, in the face of significant variations in R_s. The rotor resistance variation was successfully estimated using the adaptation capabilities of neural networks. The implementation of these techniques required only a small increase in the computation times. The feasibility and validity of the proposed identification method has been proved by the satisfactory experimental results.

It has been found that tuning a PI estimator without stability problems, is very difficult. When the PI estimator was replaced by a fuzzy logic estimator, it was easy to avoid oscillation, as the limits of the membership functions were set based on a human operator's knowledge base. The stator resistance estimator uses the error between a discrete estimate of stator current and the measured stator current, to map the change in stator resistance, using a fuzzy non-linear mapping technique. As the control surface shown in Figure 7.18 is highly non-linear, the capability of the fuzzy estimator to converge for a very wide input range is guaranteed and hence preferred with respect to the PI estimator. However, the PI estimator was very simple to implement in real time, whereas implementing the fuzzy estimator requires substantially more control resources.

It may be mentioned here that the percentage errors in estimation indicated in this chapter appears to be not significantly high but they lead to significant errors in speed estimation in the sensorless implementation of the drive as discussed in Chapter 9.

Both the PI and fuzzy estimators presented in this chapter have used the amplitude of the stator current and an initial value of stator resistance in both cases. The next chapter looks at identifying the stator resistance from the instantaneous values of stator current with the help of an ANN, which does not require any initial value.

CHAPTER 8

COMBINED ANN ON-LINE ROTOR AND STATOR RESISTANCE ESTIMATION

8.1. Introduction

In recent years, the use of Artificial Neural Networks for identification and control of nonlinear dynamic systems in power electronics and AC drives has been proposed [51],[53] and [75], because they are capable of approximating a wide range of nonlinear functions with a high degree of accuracy. In this chapter, the capability of a neural network has been deployed to have on-line estimators for both stator and rotor resistances simultaneously [76]. The proposed stator resistance observer was realized with a recurrent neural network with feedback loops trained using the standard back-propagation learning algorithm. Such an architecture is known to be a more desirable approach [77] and the implementation reported in this chapter confirms this.

The rotor resistance estimation technique has already been described in Chapter 5. The new stator resistance identification algorithm using an ANN, is described in Section 8.2. The analysis of the drive with both rotor and stator resistance estimators are investigated by modeling studies using SIMULINK and the results are discussed in Section 8.3. The satisfactory operation of the combined estimations is also tested in an experimental set-up for a squirrel-cage induction motor. These results are presented and discussed in detail in Section 8.4.

8.2. Stator resistance estimation using artificial neural networks

The voltage and current model equations of the induction motor, Equations (4.1) and (4.2) in Chapter 4, can also be written as:

$$\sigma L_s \frac{di_{ds}^s}{dt} = \frac{L_m}{L_r T_r} \lambda_{dr}^{s\ im} + \frac{L_m}{L_r} \omega_r \lambda_{qr}^{s\ im} - \frac{L_m^2}{L_r T_r} i_{ds}^s + v_{ds}^s - R_s i_{ds}^s \qquad (8.1)$$

$$\sigma L_s \frac{di_{qs}^s}{dt} = \frac{L_m}{L_r T_r} \lambda_{qr}^{s\ im} - \frac{L_m}{L_r} \omega_r \lambda_{dr}^{s\ im} - \frac{L_m^2}{L_r T_r} i_{qs}^s + v_{qs}^s - R_s i_{qs}^s \qquad (8.2)$$

Using the discrete form of (8.1),

$$i_{ds}^{s\,*}(k)=W_4 i_{ds}^{s\,*}(k-1)+W_5\lambda_{dr}^{s\,im}(k-1)+W_6\omega_r\lambda_{qr}^{s\,im}(k-1)+W_7 v_{ds}^{s}(k-1) \qquad (8.3)$$

where, $\quad W_4 =1-\dfrac{T_s}{\sigma L_s}\dfrac{L_m^2}{L_r T_r}-\dfrac{T_s}{\sigma L_s}R_s ; W_5 =\dfrac{T_s}{\sigma L_s}\dfrac{L_m}{L_r T_r}$

$W_6 =\dfrac{T_s}{\sigma L_s}\dfrac{L_m}{L_r} ; W_7 =\dfrac{T_s}{\sigma L_s}$

The weights W_5, W_6, and W_7 are calculated from the motor parameters, motor speed ω_r and the sampling interval T_s.

The relationship between stator current and stator resistance was described in Section 7.2. Their relationship is non-linear which could be easily mapped using a neural network.

Equation (8.3) can be represented by a recurrent neural network as shown in Figure 8.1. The standard back-propagation learning rule is then employed to train the network. The weight W_4 is the result of training so as to minimize the cumulative error function E_2,

$$E_2 =\frac{1}{2}\bar{\varepsilon}_2^2(k)=\frac{1}{2}\left\{i_{ds}^{s}(k)-i_{ds}^{s\,*}(k)\right\}^2 \qquad (8.4)$$

$$\Delta W_4(k)\infty-\frac{\partial E_2}{\partial W_4}=-\frac{\partial E_2}{\partial i_{ds}^{s\,*}(k)}\frac{\partial i_{ds}^{s\,*}(k)}{\partial W_4} \qquad (8.5)$$

$$-\frac{\partial E}{\partial i_{ds}^{s\,*}(k)}=\left[i_{ds}^{s}(k)-i_{ds}^{s\,*}(k)\right] \qquad (8.6)$$

$$\frac{\partial i_{ds}^{s\,*}(k)}{\partial W_4}=i_{ds}^{s}(k-1) \qquad (8.7)$$

The weight adjustment for W_4 is given by:

$$\Delta W_4(k)\infty\left[i_{ds}^{s}(k)-i_{ds}^{s\,*}(k)\right]i_{ds}^{s\,*}(k-1) \qquad (8.8)$$

To accelerate the convergence of the error back propagation learning algorithm, the current weight adjustment is supplemented with a fraction of the most recent weight adjustment, as in Equation (8.9).

$$W_4(k)=W_4(k-1)+\eta_2\Delta W_4(k)+\alpha_2\Delta W_4(k-1) \qquad (8.9)$$

where η_2 is the training coefficient, α_2 is a user-selected positive momentum constant.

Similarly, using the discrete form of (8.2),

$$i_{qs}^{s}{}^{*}(k)=W_4 i_{qs}^{s}{}^{*}(k-1)+W_5\lambda_{qr}^{s}{}^{im}(k-1)-\omega_r W_6\lambda_{dr}^{s}{}^{im}(k-1)+W_7 v_{qs}^{s}(k-1) \qquad (8.10)$$

Equation (8.7) can be represented by a neural network as shown in Figure 8.2. The weight W_4 is updated with training based on (8.9).

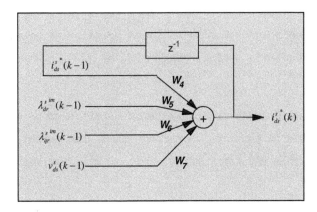

Figure 8.1 *d*-axis stator current estimation using recurrent neural network based on (8.3).

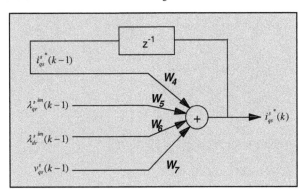

Figure 8.2 *q*-axis stator current estimation using recurrent neural network based on (8.7).

The stator resistance $\hat{R}_s$ can be calculated using Equation (8.11) as follows:

$$\hat{R}_s=\left\{1-W_4-\frac{T_s}{\sigma L_s}\frac{L_m^2\hat{R}_r}{L_r^2}\right\}\frac{\sigma L_s}{T_s} \qquad (8.11)$$

The stator resistance of an induction motor can be thus estimated from the stator current using the neural network system as indicated in Figure 8.3.

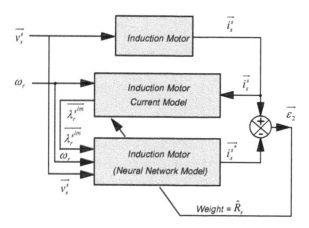

Figure 8.3 Block diagram of R_s estimation using artificial neural network.

8.3. Modeling results of rotor and stator resistance estimations using ANN

The block diagram of a rotor flux oriented induction motor drive, together with both stator and rotor resistance identifications, is shown in Figure 8.4. The induction motor is controlled with a Rotor Flux Oriented Vector Controller as shown in this figure. The voltage model fluxes are estimated from the measured stator voltages and currents using a PCLPF. The flux estimation using the PCLPF is described in Chapter 6. The stator voltages are PWM voltages and are filtered with hardware filters on the voltage sensor boards inside the IGBT inverter and only the sinusoidal voltages are taken to the PCLPF flux estimators. The rotor resistance is estimated using artificial neural networks by the *Rotor Resistance Estimator* (RRE) block. The operation of this block was described in chapter 5. The stator resistance estimation described in Section 8.2 is implemented by the *Stator Resistance Estimator* (SRE) block shown in Figure 8.4. The implementation of rotor resistance identification with an ANN supplemented with stator resistance estimation using another ANN of a type shown in Figure 8.3, has been verified by modeling studies with SIMULINK.

In order to investigate the performance of the drive for parameter variations in rotor resistance R_r, a series of simulations were conducted by introducing error between the actual value R_r and the value used in the controller R'_r. Similarly, another series of simulations were conducted by introducing error between the actual stator resistance R_s and the one used in the controller R'_s. All of these investigations were conducted for the drive running at 1000 rev/minute with a constant load torque of 7.4 Nm. The parameters of the motor used for modeling studies are shown in Table B.1.

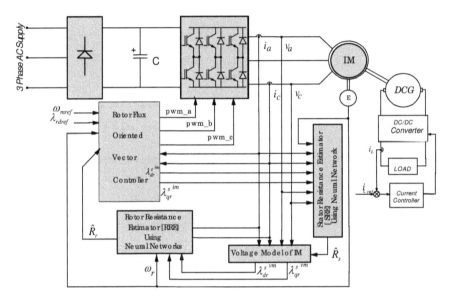

Figure 8.4 Block diagram of the RFOC induction motor drive with on-line stator and rotor resistance tracking using ANN.

8.3.1 With RRE and SRE off

Initially R_r and R_s were both increased abruptly by 40% at 1.5 second, keeping both the Rotor Resistance Estimation (RRE) and the Stator Resistance Estimation (SRE) blocks off in Figure 8.4. The steady state values of the torque, rotor flux linkage are shown in Figure 8.5. The rotor flux linkage in the motor increases by 21% compared to its estimated value, when the error in rotor resistance is introduced, as shown in Figure 8.5(iv). The estimated torque is 4% lower than the actual motor torque, as also shown in Figure 8.5(ii).

8.3.2 With RRE on and SRE off

Later, simulations were repeated after switching on only the ANN Rotor Resistance Estimation block with the SRE block still off, for the same changes introduced in Figure 8.6. The estimated $\hat{R}_r$ in this case is higher than the R_r by 1.7% as shown in Figure 8.6(i). The estimated torque is 1.35% higher than the real motor torque, as shown in Figure 8.6(ii). But the estimated rotor flux linkage is 1.5% lower than the actual rotor flux linkage as indicated in Figure 8.6(iii).

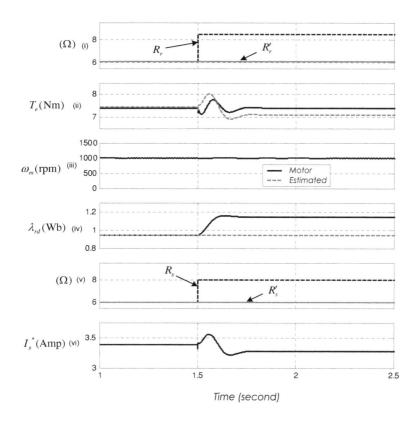

Figure 8.5 Performance of the drive without rotor and stator resistance estimations for 40 % step change in R_r and R_s, R'_r & R'_s uncompensated – modeling results.

8.3.3 With RRE on and SRE on

Finally, simulations were carried out with both the RRE and SRE blocks switched on for the same changes in R_r and R_s described previously. The results of torque, rotor flux linkage are shown for both of the cases, in Figure 8.6. The errors reported in the previous paragraph, between estimated and real quantities of torque and rotor flux linkage have largely disappeared in this case. The estimated rotor resistance has tracked the real rotor resistance of the motor very well, as the estimation error now drops to 0.3% as shown in Figure 8.6(i). However there was a 4.4% error as shown Figure 8.6(iv) for the estimated stator resistance $\hat{R}_s$ with respect to the real stator resistance R_s .

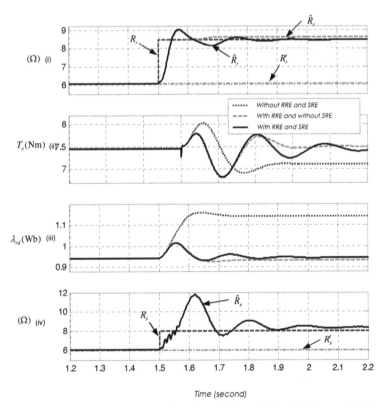

Figure 8.6 Performance of the drive with and without RRE and SRE using ANN for 40 % step change in R_r and R_s, R_r' & R_s' compensated – modeling results.

The Figures 8.5 and 8.6 described the possible steady-state errors encountered in a situation where a step change in resistance is applied, only for the purpose of investigation and verification of the technique. However, the practical variation in resistances is very slow. A corresponding modeling investigation was also carried out, and the results are indicated in Figure 8.7. The simulations are done in three steps. At first the drive system is analyzed

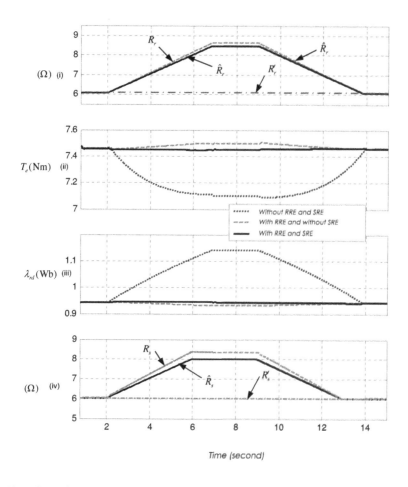

Figure 8.7 Performance of the induction motor drive for 40% ramp change in R_r and R_s with and without RRE and SRE using ANN – modeling results.

after introducing error between R_r and R'_r and R_s and R'_s keeping both the RRE and SRE turned off. Repeated simulations were also carried out, with the RRE on and SRE off. The $\hat{R}_r$ estimated in this case is higher than the R_r by 1.1% as shown in Figure 8.7(i). The estimated torque is 1.3% higher than the real motor torque, as shown in Figure 8.7(ii). However, the estimated rotor flux linkage is 1.5% lower than the actual rotor flux linkage as indicated in Figure 8.7(iii). Finally, both rotor and stator resistance estimators are investigated with both RRE and SRE switched on. The estimated rotor resistance has tracked the real rotor resistance of the motor very well, as the error now drops to 0.3% as shown in Figure 8.7(i). However there was a small but insignificant error of 5%, as shown Figure 8.7(iv), for the estimated stator resistance $(\hat{R}_s)$ with respect to the real stator resistance (R_s). However, its effect on the rotor flux oriented control is negligible, as the errors between torques and rotor flux linkages are virtually eliminated.

8.4. Experimental results of rotor and stator resistance estimations using ANN

In order to verify the proposed stator and rotor resistance estimation algorithms, the rotor flux oriented induction motor drive of Figure 8.4 was implemented in the laboratory. The details of the experimental set-up are described in Appendix-D.

In order to examine the capability of tracking the rotor resistance of the induction motor with the proposed estimator, a temperature rise test was conducted, at a motor speed of 1000 rev/min. The results of R_r and R_s estimations obtained from the experiment is shown in Figure 8.8, after logging the data for 60 minutes. Figure 8.9 shows the d–axis rotor flux linkages of the current model $(\lambda_{dr}^{s\ im})$, the voltage model $(\lambda_{dr}^{s\ vm})$ and the neural model $(\lambda_{dr}^{s\ nm})$, taken at the end of heat run. All of the flux linkages are in the stationary reference frame. The flux linkages $\lambda_{dr}^{s\ im}$ and $\lambda_{dr}^{s\ vm}$ are updated with a sampling time of 100 μsecond, whereas the flux $\lambda_{dr}^{s\ nm}$ is updated only at 1000 μsecond. The flux linkage $\lambda_{dr}^{s\ nm}$ follows the flux linkage $\lambda_{dr}^{s\ vm}$, due to the on-line training of the neural network. The coefficients used for training are, $\eta_1 = 0.005$ and $\alpha_1 = 10.0e\text{-}6$.

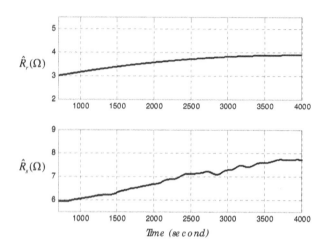

Figure 8.8 Estimation of R_r and R_s using ANN - experimental results.

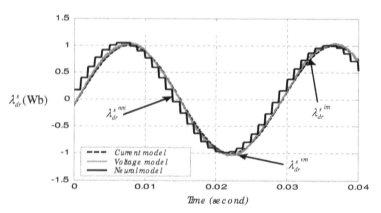

Figure 8.9 Rotor fluxes in R_r estimation - experimental results.

To test the stator resistance estimation, an additional 3.4 Ω per phase was added in series with the induction motor stator, with the motor running at 1000 rev/min and with a load torque of 7.4 Nm. The estimated stator resistance together with the actual stator resistance is shown in Figure 8.10. The estimated stator resistance converges to 9.4 Ω within less than 200 milliseconds. Figure 8.11 shows both the measured d-axis stator current and the one estimated by the neural network model. The neural model output $i_{ds}^{s*}(k)$ follows the measured values $i_{ds}^{s}(k)$, due to the on-line training of the network. The neural model current estimate is updated with a sampling time of 100μsecond. The coefficients used for training are, $\eta_2 = 0.00216$ and $\alpha_2 = 10.0e{-}6$.

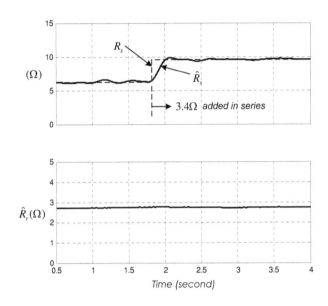

Figure 8.10 Estimated stator resistance R_s using ANN - experimental results.

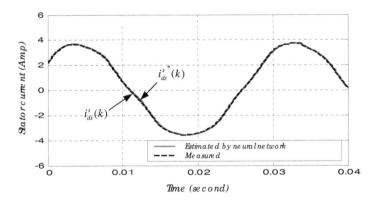

Figure 8.11 Stator currents in R_s estimation – experimental results.

8.5.   Analysis of results

The modeling results as described in Figure 8.6 indicate that the proposed rotor and stator resistance estimators can converge in a short time, as low as 200 millisecond corresponding to a 40% step change for both stator and rotor resistances simultaneously. In order to compare the stator resistance estimation for simulation and experiment, simulation is repeated with the SRE and RRE blocks on in Figure 8.4. Then a step change in R_s is applied without a step change in R_r and the results are recorded as the upper trace in Figure 8.12. The bottom trace in this figure is the same as the top trace of Figure 8.10. The estimation time in the modeled results is in very close agreement with that obtained from experiment.

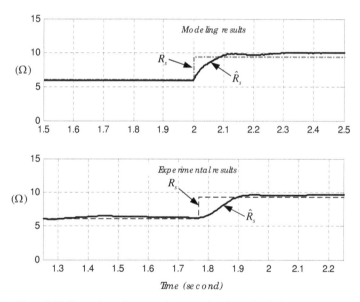

Figure 8.12 Comparison of stator resistance estimations- simulation vs. experiment.

8.6.   Conclusions

This chapter has presented a new on-line estimation technique for the rotor resistance R_r in the presence of R_s variations for the induction motor drive. The R_r estimation was found to be totally insensitive to R_s variations, as a result of the stator resistance estimation which is embedded separately.

Investigations carried out in this chapter have clearly shown that two ANNs can be used in estimating R_r in the face of significant variations in R_s, which can occur due to motor heating. Both the rotor and stator resistance variations were successfully estimated using the adaptation capabilities of neural networks. The implementation of these techniques required only a small increase of the computation times. The feasibility and validity of the proposed identification has been proved by the experimental results.

It is already established that two major parameters of the induction motor which may vary to a large extend can be estimated accurately. It is now possible to estimate the speed of the motor based on state equations, where these parameters are used. The next chapter investigates the possibility of running the motor without a speed sensor and the results are discussed.

CHAPTER 9

SPEED SENSORLESS OPERATION WITH ON-LINE ROTOR AND STATOR RESISTANCE ESTIMATIONS USING ANN

9.1. Introduction

In the preceding sections, the problem of parameter identification was analyzed or treated only for the case, where an encoder with position / speed sensing was available for use. From the investigations carried out in the previous sections, because the parameters are available with very high accuracy, it may be expected that the speed and position can also be estimated accurately. As speed sensorless operation of the induction motor plays a vital role in industry, an attempt was made to implement an RFOC induction motor drive in which the speed signal used in the RFOC is derived from the estimated fluxes in the stator reference frame. The speed sensorless operation of the induction motor drive is analyzed in this chapter.

When the speed sensor is not available, the estimation of the speed depends heavily on the rotor resistance R_r and the stator resistance R_s and these parameters depend also on the estimated speed. In the speed sensorless operation, the sampling times for the current controllers were 100 microseconds and for the speed controller was 500 microsecond in the RFOC block. The speed estimation routine was carried out every 100 microseconds. However, the estimated rotor resistance was updated only at intervals of 10 milliseconds and the estimated stator resistance was updated at 100 microseconds. Thus it was ensured that only one estimator was enabled at any point of time. In addition, the resistance estimation routines have been slowed compared to the case when the speed sensor was available. This was arranged because the estimated speed is now used in the resistance estimators instead of the measured speed. It was also assumed that the parameters are estimated only during the steady-state operation of the motor.

Both resistances are estimated experimentally in a vector controlled induction motor drive, using the method described in Chapter 8. Data on tracking performances of these estimators are presented in Section 9.3 and 9.4. With this approach the rotor resistance estimation was found to be insensitive to the stator resistance variations both in modeling and experimental results. The accuracy of the estimated speed achieved experimentally, without the speed sensor demonstrates the reliable and high performance operation of the drive [78]. The estimated and measured speeds were identical only down to the speed of 150 rpm and operation below this speed was not possible.

The rotor and stator resistance estimators are investigated by modeling studies using SIMULINK and the results are discussed in Section 9.3. The new resistance estimators are also tested in an experimental set-up for a 1.1kW squirrel-cage induction motor. The results are discussed in detail in Section 9.5.

9.2. Speed estimation

The rotor speed can be synthesized from the induction motor state equations, and can be written as:

$$\omega_r^{est} = \frac{1}{\lambda_r^2}\left[\left(\lambda_{dr}^{vm}\frac{d\lambda_{qr}^{vm}}{dt} - \lambda_{qr}^{vm}\frac{d\lambda_{dr}^{vm}}{dt}\right) - \frac{L_m}{T_r}\left(\lambda_{dr}^{vm}i_{qs} - \lambda_{qr}^{vm}i_{ds}\right)\right] \tag{9.1}$$

Equation (9.1) depends significantly on the variation of motor parameters, however as the two critical parameters for flux estimation R_r and R_s are modified by adaptation techniques, the speed estimation was found to be accurate.

9.3. Modeling results with estimated rotor speed

The block diagram of a rotor flux oriented induction motor drive, together with both stator and rotor resistance identifications, is shown in Figure 9.1. The performance of the estimators for both steady-state and dynamic situations have been investigated by the modeling studies carried out with **SIMULINK**.

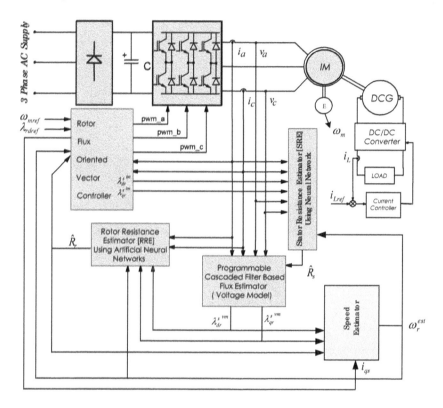

Figure 9.1 Block diagram of the speed sensorless RFOC induction motor drive with on-line stator and rotor resistance tracking.

In order to investigate the performance of the drive for parameter variations in rotor resistance R_r, a series of simulations were conducted by introducing error between the actual value R_r and the value used in the controller R'_r. Similarly, another series of simulations were conducted by introducing error between the actual stator resistance R_s and the one used in the controller R'_s. All of these investigations were conducted for the drive running at 1000 rev/minute and with a constant load torque of 7.4 Nm. The parameters of the motor used for modeling studies are shown in Table B.2. In the case of analysis with a speed sensor, discussed in Chapter 8, simultaneous abrupt changes in both R_r and R_s could be investigated. However, in the speed sensorless analysis, this analysis was not possible. However, the method can perform the estimations at reduced execution times as described in 9.1, where the resistance variation is only very slow. A corresponding investigation was carried out, and the results are indicated in Figure 9.2. The simulations are done in three steps.

9.3.1 With RRE and SRE off

At first the drive system was analyzed after introducing error between R_r and R'_r and R_s and R'_s keeping both the rotor resistance (RRE) and the stator resistance (SRE) estimators in the Figure 9.1 disabled. The rotor resistance R_r was increased from 6.03 Ω to 8.5 Ω over 20 seconds and kept constant for more than 15 seconds and then gradually decreased back to 6.03 Ω over 20 seconds as shown in Figure 9.2(i). The stator resistance R_s was increased from 6.03 Ω to 8.0 Ω over 20 seconds and kept constant for more than 15 seconds and then gradually decreased back to 6.03 Ω over 20 seconds as shown in Figure 9.2(iv). The estimated torque is 4% lower than the actual motor torque as shown in Figure 9.2(ii). The estimated rotor flux linkage λ_{rd} has increased by 21% as indicated in Figure 9.2(iii). This increase of flux linkage λ_{rd} is based on the assumption of a linear magnetic circuit.

9.3.2 With RRE on and SRE off

Subsequently, simulation was carried out after enabling the RRE block in Figure 9.1, keeping the stator resistance estimator (SRE) disabled. The $\hat{R}_r$ estimated in this case is higher than the actual R_r by 1.7% as shown in Figure 9.2(i). The estimated torque is 1.35% higher than the real motor torque, as shown in Figure 9.2(ii). However, the estimated rotor flux linkage is 1.5% lower than the actual rotor flux linkage as indicated in Figure 9.2(iii).

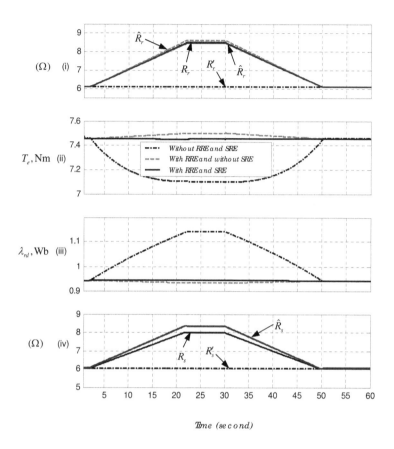

Figure 9.2 Performance of the speed sensorless induction motor drive for a 40% ramp change in R_s and R_r with and without RRE and SRE using ANN - modeling results.

9.3.3 With RRE on and SRE on

Finally, the simulation was carried out after enabling both the rotor and stator resistance estimators. The dynamic torque and rotor flux linkage are shown for both of the cases, in Figure 9.2. The error reported in the previous paragraph, between estimated and actual motor torques and rotor flux linkages have largely disappeared in this case. The estimated rotor resistance has tracked the actual rotor resistance of the motor very well, as the estimation error now drops to 0.3% as in Figure 9.2(i).

However it can be noted that there is a 4.4% error for the estimated stator resistance with respect to the actual stator resistance R_s as shown Figure 9.2(iv). However, its effect on the rotor flux oriented control is negligible, as the errors between torques and rotor flux linkages are virtually eliminated.

9.4. Experimental results with estimated rotor speed

In order to verify the tracking of stator and rotor resistance estimations in the speed sensorless condition, a rotor flux oriented induction motor drive was implemented in the laboratory. The experimental set-up is discussed in detail in Appendix -D.

To examine the capability of tracking the rotor and stator resistance of the induction motor with the proposed estimator, a temperature rise test was conducted, at a motor speed of 1000 rev/min. The results of estimated rotor resistance $\hat{R}_r$ obtained from the experiment are shown in Figure 9.3, after logging the data for 60 minutes. Figure 9.4 shows the d–axis rotor flux linkages of the current model ($\lambda_{dr}^{s\ lm}$), the voltage model ($\lambda_{dr}^{s\ vm}$) and the neural model ($\lambda_{dr}^{s\ nm}$), taken at the end of heat run. All of these flux linkages are in the stationary reference frame. The flux linkages $\lambda_{dr}^{s\ lm}$ and $\lambda_{dr}^{s\ vm}$ are

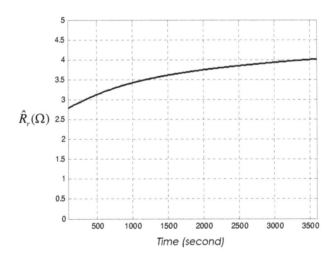

Figure 9.3 Estimated resistance $\hat{R}_r$ without a speed sensor - experimental results.

updated with a sampling time of 100 µsecond, whereas the flux $\lambda_{dr}^{s\;nm}$ is updated only at 1000 µsecond. The flux linkage $\lambda_{dr}^{s\;nm}$ follows the flux linkage $\lambda_{dr}^{s\;vm}$, due to the on-line training of the neural network. The coefficients used for training are, $\eta_1 = 0.005$ and $\alpha_1 = 10.0\text{e-}6$.

To verify the stator resistance estimation, an additional 3.4 Ω per phase was added in series with the induction motor stator, with the motor running at 1000 rev/min and with a load torque of 7.4 Nm. The estimated stator resistance together with the actual stator resistance is shown in Figure 9.5. The estimated stator resistance converges to 9.4 Ω within less than 200 millisecond. Figure 9.6 shows both the measured d-axis stator current and the one estimated by the neural network model. The neural model output $i_{ds}^*(k)$ follows the measured values $i_{ds}(k)$, due to the on-line training of the network. The neural model current estimate is updated with a sampling time of 100 µsecond. The coefficients used for training are, $\eta_2 = 0.00216$ and $\alpha_2 = 10.0\text{e-}6$.

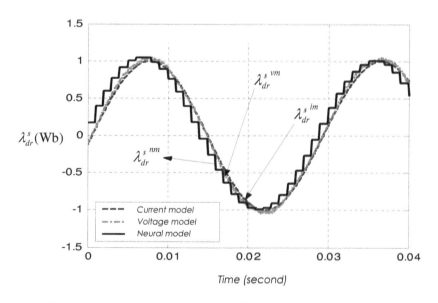

Figure 9.4 Rotor fluxes during R_r estimation without a speed sensor - experimental results.

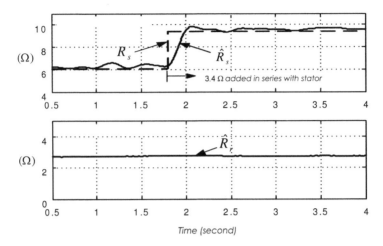

Figure 9.5 Estimated stator resistance $\hat{R}_s$ for a step change in R_s without a speed sensor
- experimental results.

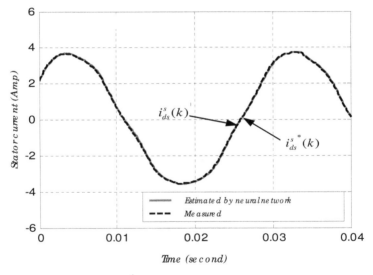

Figure 9.6 Stator currents in $\hat{R}_s$ estimation without a speed sensor - experimental results.

Experiments were conducted to verify the performance of the speed sensorless operation of the drive together with the proposed rotor and stator resistance adaptation. To examine the effect of stator resistance in the estimated speed, an additional 3.4 Ω per phase was added in series with the induction motor stator, with the drive operated with RFOC and with the SRE block in Fig. 9.1 off. The estimated speed dropped by 10 rev/min from 1000 rev/min as indicated in Fig. 9.7. Fig. 9.8 shows the effect of stator resistance estimation during the speed reversal. Fig. 9.8(i) shows the plots of estimated speed ω_r^{em} and the speed ω_r measured by an incremental encoder, before switching in the additional 3.4 Ω resistance. Fig. 9.8(ii) shows the results after switching in the additional resistance with the SRE block off. Finally, the SRE block is turned on and the results are recorded as in Fig. 9.8(iii). The distortion in the estimated speed in Fig. 9.8(ii) has totally disappeared in Fig. 9.8(iii). An additional detailed view of the speed estimation results of Fig. 9.8(iii) is plotted separately in Fig. 9.9 for better readability.

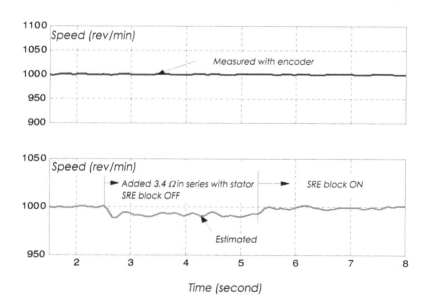

Figure 9.7 Effect of stator resistance estimator on steady state speed estimation
- experimental results.

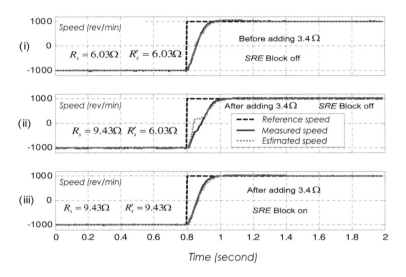

Figure 9.8 Effect of stator resistance estimation on dynamic speed estimation
- experimental results.

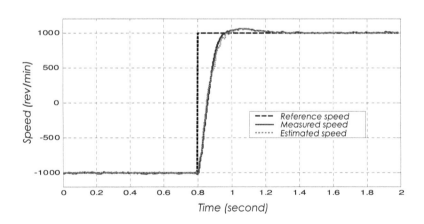

Figure 9.9 Dynamic speed estimation - experimental results.

In order to examine the capability of the speed sensorless drive for load transients, a step load of 7.4 Nm was applied to the motor shaft, and the estimated and measured speeds were monitored. Fig. 9.10 shows these results when the drive was operating at 1000 rev/min.

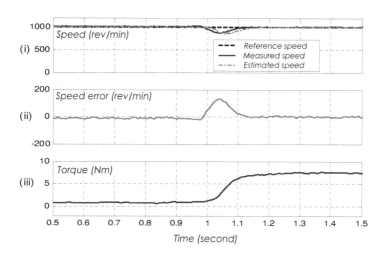

Figure 9.10 Speed estimation during a step load - experimental results.

To analyze the low speed performance of the RFOC drive with the the resistance estimators, the experiment was repeated for a speed reversal from -150 rev/min to +150 rev/min and the speed estimation was found to be satisfactory as shown in Fig. 9.11.

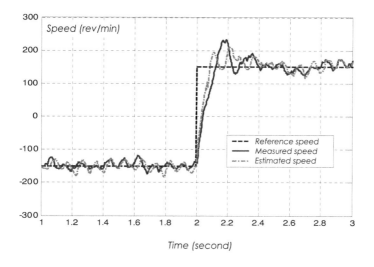

Figure 9.11 Experimental results for low speed estimation.

9.5. Conclusions

In this chapter, the rotor resistance R_r and stator resistance R_s are estimated using an ANN without using a speed sensor. This approach is in contrast with the methods in chapter 8 where measured speed was used in the current model in estimating the rotor flux. In this chapter, the current model utilizes speed which is also estimated. This was possible by executing the speed estimation algorithm much faster than the resistance estimation algorithms. The underlying assumption is that speed remains constant during flux estimation and rotor resistance remains constant during speed estimation.

It has been shown resistance estimation in this speed sensorless approach is still as accurate as in chapter 8. The convergence of the stator resistance estimator is evident from the tracking of the estimated stator current with the measured stator current as shown in Figure 9.6. The estimated stator resistance was also able to converge for a step change in stator resistance in the absence of a speed sensor as shown in Figure 9.5, just as it did in the case with the speed sensor presented in Figure 8.10.

The estimated speed was verified with the measured speed using an encoder. They were compared and found to be in agreement during transient condition as shown in Figure 9.9 and during steady state as shown in Figure 9.10.

CHAPTER 10

CONCLUSIONS

10.1 Conclusions of this book

This book reports the results of a detailed investigation of the application of artificial neural networks, fuzzy and other estimators in tracking the rotor resistance of rotor flux oriented induction motor drives. The main aim of the project was to track the rotor resistance of a rotor flux oriented induction motor drive quickly and accurately, using artificial neural networks. The critical parameter in a rotor flux oriented induction motor drive is the rotor resistance. The problem of rotor time constant or rotor resistance adaptation in rotor flux oriented vector control of induction motors has been a subject of vigorous research for the last twenty years. The ANN and fuzzy logic techniques for on-line estimation have also been receiving considerable interest during this period, because of some of their advantages over classical methods. This book has included a review of the literature on the effects of rotor resistance variation for the RFOC drive and the techniques of applying ANN and fuzzy estimators, in chapters 2 and 3 respectively.

In order to conduct a thorough investigation of the performance of a rotor flux oriented induction motor drive with rotor resistance tracking by ANN and other methods, a mathematical model of the basic RFOC drive with a PWM inverter was developed using SIMULINK. All the additional identification blocks were then added to this basic RFOC system. To validate the mathematical models developed, an experimental RFOC drive was set up with both slip-ring and squirrel-cage induction motors. The performances of these drive systems in simulation and experiment were found to be in good agreement, as detailed in chapter 4. The simulation and the experimental systems then provided the platform for understanding the problems of parameter detuning in the RFOC and the estimation and adaptation methods developed in this project.

A simple PI and a fuzzy logic controller were then tested to adjust the rotor resistance in the RFOC drive. It was shown that such estimators could not track the rotor resistance during the speed and load torque dynamics of the drive, satisfactorily.

Preliminary investigations carried out with *off-line* training of a neural network for estimating the motor speed was found to be unsatisfactory. Further investigations then focused only on *on-line* learning techniques, as their learning performance was found to be much superior to off-line training methods. The rotor resistance identification finally adopted is similar to that of a model reference adaptive system using voltage and current models of the induction motor, except for the fact that the motor current model was replaced with an ANN based model. As the adaptation was realized with neural network models, adaptation based on instantaneous induction motor states was made, resulting in a very fast and accurate convergence of the estimation procedure, as shown in chapter 5.

The proposed rotor resistance estimator using an ANN for a 3.6 kW slip-ring induction motor was validated via modeling and experimental measurements. The error of the rotor resistance estimation for this machine was found to be no more than 4%, over the expected range of resistance variation of the motor. Subsequently further modeling and experiments were repeated with a 1.1kW squirrel-cage induction motor and was found to give very good results.

In this book, it has been established that a programmable cascaded filter could be effectively used to remove the influence of the measurement offsets and estimate the stator flux from the machine back-emf. The rotor flux linkages needed for rotor resistance estimation was then calculated by using the stator flux from the cascaded filter and the stator current. The modeling and experimental results presented in chapter 6 clearly show that the flux estimation results of the PCLPF method were in very good agreement with the rotor flux estimated by the current model during both dynamic and steady-state operation of the RFOC drive. The PCLPF did not require any arbitrary selection of gains or tuning and it was very simple to be implemented in real time.

The rotor resistance identification technique of chapter 5 was partly dependent on the stator resistance of the induction motor. PI, fuzzy and ANN based stator resistance estimators were tested. It was found that the PI estimator was highly susceptible to controller gains. It was found that fuzzy estimator for tracking R_s had no stability issues. However both the PI and fuzzy estimators could not give fast estimation because they depended on the amplitudes of measured and estimated currents.

A new ANN based stator resistance estimator was then developed which largely eliminated the problems encountered with the PI and fuzzy stator resistance estimators, as presented in chapter 8. Modeling results were compared with the results taken from the experimental set-up. They were found to be in very close agreement. By including this ANN stator resistance estimator along with the ANN rotor resistor estimator, the final estimation of rotor resistance was further improved.

The studies on the development of ANN and other estimators initially assumed the availability of an accurate speed sensor. Because speed sensorless operation of drives has recently become an important research goal in its own right, an attempt was made to include a speed estimator with the other estimators. The rotor speed was estimated using the simple state equations of the induction motor. As the parameters of the induction motor were already tracked accurately, the estimated speed based on the state estimation was found accurate enough for the closed loop operation of the drive. It was found from the results presented in chapter 9 that the estimators discussed in chapter 8 could also operate with the estimated speed. However, the minimum speed achievable was 150 rev/min, in the speed sensorless drive. Further investigations are necessary to improve the performance of the speed sensorless drive at speeds below 150 rev/min.

10.2 Suggestions for future work

When the rotor resistance was estimated in the experiment with the slip-ring induction motor, the estimated rotor resistance was found to have an error of nearly 4%. Part of this error could be attributable to brush resistance and brush contact drops of the slip-ring motor, which were ignored in both modelling and experimental studies. This aspect could be investigated further.

The sampling interval for the fastest algorithms used in this project could not be set below 100 microseconds. The sampling time is an issue relative to cost and the DSP processor system employed. Faster DSPs may reduce this limitation in future. The computations within ANN estimators are assumed to happen in parallel. However, when these estimators are implemented with a sequential computation platform like a DSP the best results expected with ANN are not guaranteed. One of the simple solutions to this problem is to reduce the execution time of these estimators to as low as possible. Further investigations are required with a faster sampling frequency so as to resemble parallel computation.

The measured motor terminal voltages were the PWM voltages which were filtered before feedback, thus introducing a small delay and attenuation. The experimental results discussed in this book have used a PWM inverter with a switching frequency of 5 kHz. The motor terminal voltages are measured directly across the PWM inverter terminals and these signals are filtered using hardware filters so that they could be sampled into the DSP system. This method of measurement leads to errors especially at lower speeds. If space vector modulation with DC link voltage measurement and dead time compensation is used, the abovementioned problems associated with filtering the PWM voltage signals from machine terminals could be reduced.

As noted in Chapters 6 and 8, the rotor resistance identification was carried out with a feedforward neural network without feedback and the stator resistance identification with a recurrent neural network which has a feedback input. It is felt that the learning capability of the rotor resistance estimator could still be improved by modifying the network with recurrent neural networks. The ANN based rotor resistance estimation structures with feedback input may be investigated in future.

This book primarily focussed on the design of the flux and resistance estimators. It was shown that the resistances are estimated accurately. The consequent enhancements in the dynamic performances of the drive may be further studied.

REFERENCES

[1] F. Blaschke, "The principle of field orientation as applied to the new transvector closed loop system for rotating field machines," *Siemens Review,* vol. 34, pp. 217–220, May 1972.

[2] M. Depenbrock, "Direct self control (DSC) of inverterc fed induction machine," *IEEE Transactions on Power Electronics*, vol.3, pp. 420-429, Oct. 1988.

[3] I. Takahashi and T. Noguchi, "A new quick torque response and high efficiency control strategy of an induction motor," *IEEE Transactions on Industry Applications*, vol. 22, pp. 820-827, Sept./Oct. 1986.

[4] B.K. Bose, *Modern Power Electronics and AC Drives*, Prentice Hall, New Jersey, 2002.

[5] E. Ho and P.C. Sen, "Decoupling control of induction motor drives", *IEEE Transactions on Industrial Electronics*, vol. 35, No.2, pp. 253-262, 1988.

[6] K. Hasse, "Zur dynamik drehzahlgeregelter antriebe mit stromrichtergespeisten asynchron-kurzschlusslaufermaschinen," *Ph.D. dissertation*, Techn. Hochsch., Darmstadt, Germany, 1969.

[7] R.Gabriel, W.Leonard and C. Nordby, "Field oriented control of standard AC motor using microprocessor," IEEE Transactions on Industry Applications, vol. IA-16, no.2, 1980, pp. 186-192.

[8] P.L. Jansen and R.D. Lorenz, "A physically insightful Approach to the Design and Accuracy Assessment of Flux observers for Field Oriented Induction Machine Drives", *IEEE Transactions on Industry Applications*, vol. 30, no.1, pp. 101-110, Jan/Feb 1994.

[9] P.L. Jansen, R.D. Lorenz and D.W. Novotny, "Observer-based direct field orientaion: Analysis and comparison of alternate methods," *IEEE Transactions on Industry Applications*, vol. 30, no. 4, pp.945-953,July./Aug. 1994.

[10] Z. Yan, C. Jin and V. Utkin, "Sensorless sliding mode control of induction motors," *IEEE Transactions on Industrial Electronics*, vol. 47, no.6, Dec 2000, pp. 1286-1297.

[11] A. Benchaib, A. Rachid, E. Auderzet and M. Tadjine, "Real-time sliding-mode observer and control of an induction motor," IEEE Transactions on Industrial Electronics, vol. 46, no.1, pp. 128-138, Feb. 1999.

[12] H. Rehman, A. Derdiyok, M.K. Guven and L. Xu, "A new current model flux observer for wide speed range sensorless control of an induction machine," IEEE Transactions on Power Electronics, vol. 17, no. 6, Nov. 2002, pp. 1041-1048.

[13] R. Krishnan and F.C. Doran, "Study of parameter sensitivity in high-performance inverter-fed induction motor drive systems," IEEE Transactions on Industry Applications, vol. IA-23, no. 4, July. /Aug. 1987, pp. 623-635.

[14] T. Matsuo, T.A. Lipo, "A rotor parameter identification scheme for vector controlled induction motor drives," *IEEE Transactions on Industry Appilcations*, vol.21, pp. 624-632, May /June 1985.

[15] H.A. Toliyat and A.A.GH. Hosseiny, "Parameter estimation algorithm using spectral analysis for vector controlled induction motor drives," in *Proc. IEEE Interanational Symposium on Industrial Electronics*, 1993, pp. 90-95.

[16] R. Gabriel and W. Leonard, "Microprocessor control of induction motor," in *Proc. International Semiconductor Power Conversion Conference*, 1982, pp. 385-396.

[17] H. Sugimoto and S. Tamai, "Secondary resistance identification of an induction motor applied model reference adaptive system and its characteristics," *IEEE Transactions on Industry Appilcations*, vol. IA 23, pp. 296-303, Mar. /Apr. 1987.

[18] L.C. Lai, C.L. Demarco and T.A. Lipo, "An extended kalman filter approach to rotor time constant measurement in PWM induction motor drives," *IEEE Transactions on Industry Appilcations*, vol. 28, pp. 96-104, Jan. /Feb. 1992.

[19]   J.W. Finch, D.J. Atkinson and P.P. Acarnley, "Full order estimator for induction motor states and parameters," *Proc. Institute of Electrical Engineers, Electric Power Applications*, vol. 145, no. 3, pp. 169-179, 1998.

[20]   R. Krishnan and F.C. Doran, "A method of sensing line voltages for parameter adaptation of inverter-fed induction motor servo drives," in *Proc. IEEE Industry Applications Society Annual Meeting*, 1985, pp. 570-577.

[21]   L.J. Garces, "Parameter adaptation for the speed-controlled static ac drive with a squirrel cage induction motor," *IEEE Transactions on Industry Appilcations*, vol. 16, pp. 173-178, Mar. /Apr. 1980.

[22]   T. Rowan, R. Kerkman and D. Leggate, "A simple on-line adaptation for indirect field orientation of an induction machine," *IEEE Transactiomns on Industry Applications*, vol.37, pp.720-727, July / Aug.1991.

[23]   R. D. Lorenz and D. B. Lawson, "A Simplified approach to continuous on-line tuning of field oriented Induction motor Drives," *IEEE Transactions on Industry Applications*, vol.26, No.3, pp.420-424, May/June 1990.

[24]   D. Dalal and R. Krishnan, "Parameter compensation of indirect vector controlled induction motor drive using estimated airgap power," *Proc. IEEE Industry Applications Society Annual Meeting, IAS1987*, pp. 170-176.

[25]   L.Umanand and S.R. Bhat, "Adaptation of the rotor time constant for variations in the rotor resistance of an induction motor," *Proc. IEEE Power Electronics Specialists Conference PESC'94*, pp. 738-743.

[26]   F.Loeser and P.K. Sattler, "Identification and compensation of the rotor temperature of AC Drives by an observer," *IEEE Transactions on Industry Applications*, vol.21, pp. 1387-1393, Nov./Dec. 1985.

[27]   M.P.Kazmierkowski and W. Sulkowski, "Transistor inverter-fed induction motor drive with vector control system," *Proc. IEEE Industry Applications Society Annual Meeting, 1986*, pp. 162-168.

[28] C. C. Chan and H. Wang, "An effective method of rotor resistance identification for high performance induction motor vector control," *IEEE Transactions on Industrial Electronics*, vol. 37, pp. 477-482, Dec. 1990.

[29] H. Toliyat, M.S. Arefeen, K.M. Rahman, and M. Eshani, "Rotor time constant updating scheme for a rotor flux oriented induction motor drive," *IEEE Transactionson Power Electronics*, vol. 14, pp. 850-857, Sept. 1999.

[30] F. Zidani, M.S. Nait-Said, M.E.H. Benbouzid, D. Diallo and R. Abdessemed, "A fuzzy rotor resistance updating scheme for an IFOC induction motor drive," *IEEE Power Engineering Review*, pp. 47-50, Nov. 2001.

[31] E. Bim, "Fuzzy optimization for rotor constant identification of an indirect FOC induction motor drive," *IEEE Transactions on Industrial Electronics*, vol. 48, pp. 1293-1295, Dec. 2001.

[32] M. Ta-Cao and H. Le-Huy, "Rotor resistance estimation using fuzzy logic for high performance induction motor drives," in *Proc. IEEE Industrial Electronics Society Annual Meeting, 1998, pp. 303-308.*

[33] D. Fodor, G. Griva and F. Profumo, "Compensation of parameters variations in induction motor drives using a neural network," in *Proc. IEEE Power Electronics Specialists Conference, 1995, pp. 1307-1311.*

[34] A. Ba-Razzouk, A. Cheriti and G. Olivier, "Artificial neural networks rotor time constant adaptation in indirect field oriented control drives," in *Proc. IEEE Power Electronics Specialists Conference, 1996, pp. 701-707.*

[35] S. Mayaleh and N.S. Bayinder, "On-line estimation of rotor time constant of an induction motor using recurrent neural networks," in *Proc. IEEE Workshop Computers in Power Electronics, 1998, pp. 219-223.*

[36] R.J. Kerkman, B.J. Seibel, T.M. Rowan and D.W. Schlegel, "A new flux and stator resistance identifier for AC Drive systems," *IEEE Transactions on Industry Applications*, vol. 32, pp. 585-593, May / June 1996.

[37] T.G. Habetler, F. Profumo, G. Griva, M. Pastorelli and A. Bettini, "Stator resistance tuning in a stator-flux field-oriented drive using an instantaneous

hybrid flux estimator," *IEEE Transactions on Power Electronics*, vol. 13, pp. 125-133, Jan. 1998.

[38]   R. Marino, S. Peresada, and P.Tomei, "On-line stator and rotor resistance estimation for induction motors," *IEEE Transactions on Control System Technology*, vol. 8, pp. 570-579, May 2000.

[39]   B.K. Bose and N.R. Patel, "Quassy-fuzzy estimation of stator resistance of induction motor," *IEEE Transactions on Power Electronics*, vol. 13, pp. 401-409, May 1998.

[40]   G. Guidi and H. Umida, "A novel stator resistance estimation method for speed-sensorless induction motor drives," *IEEE Transactions on Industry Applications*, vol. 36, pp.1619-1627, Nov. / Dec. 2000.

[41]   H. Tajima, G. Guidi and H. Umida, "Consideration about problems and solutions of speed estimation method and parameter tuning for speed-sensorless vector control of induction motor drives," *IEEE Transactions on Industry Applications*, vol. 38, pp. 1282-1289, Sept. / Oct. 2002.

[42]   J. Holtz and J. Quan, "Sensorless vector control of induction motors at very low speed using a nonlinear inverter model and parameter estimation," *IEEE Transactions on Industry Applications*, vol. 38, pp. 1087-1095, July / Aug. 2002.

[43]   K. Akatsu and A. Kawamura, "Sensorless very low-speed and zero-speed estimations with online rotor resistance estimation of induction motor without signal injection," *IEEE Transactions on Industry Applications*, vol. 36, pp. 764-771, May / June 2000.

[44]   I. Ha and S.H. Lee, "An online identification method for both stator and rotor resistances of induction motors without rotational transducers," *IEEE Transactions on Industrial Electronics*, vol. 47, pp. 842-853, Aug. 2000.

[45]   P. Vas , *Artificial -Intelligence-Based Electrical Machines and Drives: application of fuzzy, neural, fuzzy-neural and genetic-algorithm – based techniques*, Oxford University Press, New York ,1999.

[46] *Neural Network Toolbox User's Guide for use with Matlab*, Mathworks Inc, Version 3, 1998.

[47] C. Cybenko, "Approximations by superposition of a sigmoidal function," *Mathematics, Control, Signal, Systems*, vol. 2, 1989, pp. 303-314.

[48] B.K.Bose, "Expert System, Fuzzy Logic, and Neural Network Applications in Power Electronics and Motion Control", *Proceedings of the IEEE*, vol.82, No.8, August 1994.

[49] A.K.P. Toh, E.P. Nowicki and F.A. Ashrafzadeh, "A flux estimator for field oriented control of an induction motor using an artificial neural network," *Conference Rec. IEEE Industry Applications Society Annual Meeting*, Oct. 1994 pp. 585 – 592.

[50] M. Mohamadian, E.P. Nowicki and J.C. Salmon, "A neural network controller for indirect field orientation control," *Conference Rec. IEEE Industry Applications Society Annual Meeting* , Oct. 1995, pp. 1770-1774.

[51] M.G. Simoes anfd B.K. Bose, "Neural network based estimator of feedback signals for a vector controlled induction motor drive," *IEEE Transactions on Industry Applications*, vol. 31, pp. 620-629, May / June 1995.

[52] A. Ba-Razzouk, A. Cheriti, G. Oliver and P. Sicard, "Field oriented control of induction motors using neural-network decouplers," *IEEE Transactions on Power Electronics*, vol. 12, pp. 752-763, July 1997.

[53] L.Ben-Brahim, S. Tadakuma and A. Akdag, "Speed control of induction motor without rotational transducers," *IEEE Transactions on Industry Applications,* vol.35, no.4, pp.844 – 850, July/August 1999.

[54] S.H Kim, T.S Park, J.Y. Yoo and G.T Park, "Speed-sensorless vector control of an induction motor using neural network speed estimation," *IEEE Transactions on Industrial Electronics*, vol. 48, pp. 609-614, June 2001.

[55] G.C.D. Sousa, B.K. Bose and K.S. Kim, "Fuzzy logic based on-line MRAC tuning of slip gain for an indirect vector controlled induction motor drive,"

Conference Rec., Annual Conference of the IEEE Industrial Electronics Society, IECON 1993 pp. 1003-1008.

[56] G.C.D. Sousa, B.K. Bose and J.G. Cleland, "Fuzzy Logic based on-line efficiency optimisation control of an indirect vector controlled induction motor drive," *IEEE Transactions on Industrial Electronics.,* vol. 42, no.2, pp. 192-198, April 1995.

[57] B. Karanayil, M. F. Rahman and C. Grantham, "A complete dynamic model for a PWM VSI-fed rotor flux oriented vector controlled induction motor drive using SIMULINK," *Proc. International Power Electronics and Motion Control Conference, PEMC'2000*, August 2000, vol. 1, pp 284-288.

[58] R. Krishnan and A. S. Bharadwaj, "A review of parameter sensitivity and adaptation in indirect vector controlled induction motor drive systems", *IEEE Transactions on Power Electronics*, vol.6, no. 4, pp.623-635, Oct 1991.

[59] B. Karanayil, M. F. Rahman and C. Grantham, "PI and Fuzzy estimators for on-line tracking of rotor resistance of indirect vector controlled Induction motor drive", *Proc. International Electrical Machines and Drives Conference, IEMDC 2001*, June 2001, pp. 820-825.

[60] J. Yen and R. Langari, *Fuzzy Logic: Intelligence, Control, and Information*, Prentice Hall, New Jersey, 1999.

[61] *FuzzyTECH5.3* User's Manual, Inform Software Corp, 1999.

[62] K. Funahashi, "On the approximate realization of continuous mappings by neural networks," *Neural Networks*, vol.2, pp. 183-192, 1989.

[63] K. Hornik, M. Stinchcombe and H. White, "Multilayer feedforward networks are universal approximations," *Neural Networks*, vol.2, pp.359-366, 1989.

[64] B. Karanayil, M. F. Rahman and C. Grantham, "Rotor resistance identification using artificial neural networks for an indirect vector controlled induction motor drive", *Proc. Annual Conference of Industrial Electronics Society, IECON2001*, November 2001, vol. 2, pp. 1315-1320.

[65] M. Hinkkanen and J. Luomi, "Modified Integrator for Voltage Model Flux Estimation of Induction Motors", *Proc. Annual Conference of the IEEE Industrial Electronics Society, IECON 2001,* Nov. 2001, vol. 2, pp. 1339-1343.

[66] J. Hu and B. Wu, "New Integration algorithms for estimating motor flux over a wide speed range", *IEEE Transactions on Power Electronics,* vol. 13, No. 5, pp. 969-977, Sep. 1998.

[67] B.K. Bose and N.R. Patel, "A Programmable Cascaded Low Pass Filter based flux synthesis for a Stator Flux Oriented vector controlled induction motor drive", *IEEE Transactions on Industrial Electronics,* vol.44, No. 1, pp. 140-143, Feb 1997.

[68] M.E. Haque, L.Zhong and M.F. Rahman, "The effect of offset error and its compensation for a direct torque controlled interior permanent magnat synchronous motor drive," *Proc. International Electrical Machines and Drives Conference, IEMDC 2001,* June 2001, pp. 814-819.

[69] G.C. Verghese and S.R. Sanders, "Observers for Flux Estimation in Induction Machines", *IEEE Transactions on Industrial Electronics,* vol.35, No.1, pp. 85-94, Feb 1988.

[70] H. Rehman, M.K. Guven, A. Derdiyok and L. Xu, "A new current model flux observer insensitive to rotor time constant and rotor speed for DFO control of induction machine", *Proc. IEEE Power Electronics Specialists Conference,* June 2001, vol.2, pp. 1179 –1184.

[71] J.H. Kim, J.W. Choi, S.K. Sul, "Novel rotor flux observer using observer characteristic function in complex vector space for field oriented induction motor drives", *Proc. IEEE Applied Power Electronics Conference,* March 2001, vol. 1, pp. 615 –621.

[72] H. Kubota, K. Matsuse and T. Nakano, "New adaptive flux observer of induction motor for wide speed range motor drives", *Proc. Annual Conference of the IEEE Industrial Electronics Society, IECON'90,* pp. 921-926.

[73] B. Karanayil, M.F. Rahman and C. Grantham, "An implementation of a Programmable Cascaded Low-Pass Filter for a rotor flux synthesizer for an induction motor drive", *IEEE Transactions on Power Electronics*, vol. 19, No.2, pp. 257-263, March 2004.

[74] B. Karanayil, M. F. Rahman and C. Grantham, "Stator and rotor resistance observers for induction motor drive using Fuzzy Logic and Artificial Neural Networks", *Proc. IEEE Industry Applications Society Annual Meeting, IAS2003,* October 2003, pp.124-131.

[75] S.K. Mondal, J.O.P. Pinto and B.K. Bose, " A Neural network based space-vector PWM controller for a three voltage-fed innverter induction motor drive," *IEEE Transactions on Industry Applications*, vol.38, No.3, pp.660-669, May 2002.

[76] B. Karanayil, M. F. Rahman and C. Grantham, "On –line stator and rotor resistance estimation scheme for vector controlled induction motor drive using Artificial Neural Networks", *Proc. IEEE Industry Applications Society Annual Meeting, IAS2003,* October 2003, pp. 132-139.

[77] D.T. Pham and X. Liu, *Neural Networks for Identification, Prediction and Control*, Springer –Verlag, NewYork, 1995.

[78] B. Karanayil, M.F. Rahman and C. Grantham, "Rotor resistance identification using Artificial Neural Networks for a speed sensorless vector controlled induction motor drive", *Proc. Annual Conference of the IEEE Industrial Electronics Society, IECON 2003*, November 2003, pp. 419-424.

[79] K.S. Narendra and K. Parthasarathy, "Identification and control of dynamical systems using neural networks," *IEEE Transactions on Neural Networks.*, vol. 1, no.1, pp. 4-27, Mar. 1990.

[80] S. Haykin, *A Comprehensive Foundation*, Macmillan College Publishing Company, 1994.

[81] C.T. Lin and G. Lee, *Neural Fuzzy Systems – A Neuro Fuzzy Synergism to Intelligent Systems*, Prentice Hall PTR, Upper Saddle River, NJ 07458, 1996.

[82]   L.A. Zadeh, *"Fuzzy Sets,"* *Information and Control*, vol. 8, pp. 338-353, 1965.

[83]   G.J. Klir and B. Yuan, *Fuzzy sets and fuzzy logic*, Prentice Hall, 1995.

[84]   S. Assilian and E. Mamdani, "An experiment in linguistic synthesis with a fuzzy logic controller," *International Journal of Man-Machine Studies*, vol. 7, pp. 1-3, 1975.

[85]   D. Driankov, H. Hellendoorn and M. Reinfrank, *An Introduction to Fuzzy Control*, Springer-Verlag, Berlin, 1993.

APPENDIX A

INTRODUCTION TO ARTIFICIAL NEURAL NETWORKS AND FUZZY LOGIC

A.1 Introduction to Artificial Neural Networks

Artificial Neural Networks have emerged as a powerful problem solving tool in the area of pattern recognition and function emulation. There are various types of ANNs though they can be broadly classified into three main categories – Feedforward, Feedback and self-organising. Two classes of neural networks which have received considerable attention in the area of identification and control of dynamic systems are:

- Multilayer feedforward neural networks
- Recurrent networks, which can be viewed as generalized feedforward networks with some of the outputs being used as inputs after delay [79].

The most common type is known as the multilayer feedforward network. The basic structure of a multilayer feedforward network is shown in Figure A.1. Feedforward networks are so named because the output of each layer feeds the next layer of units. The Perceptron, proposed by Rosenblatt in 1962 and Adaline proposed by Widrow in the same year are the earliest feedforward architectures. These ANNs consist of two

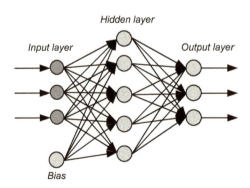

Figure A.1 Three layer feedforward neural network.

basic parts – a typically non-linear processing element called the neuron, and a connection element called the synapse which connects various neurons. Each synapse has a number associated with it, called the synaptic weight. All the knowledge in the ANN is stored in these weights, also known as free parameters of the network.

Figure A.2 shows one form of a recurrent network known as a Jordan network, where some of the outputs are fed back as inputs to the network through a delay element. Many other forms of recurrent networks have been proposed including Elman networks, where the hidden neuron outputs are fed back to themselves through a delay. While feedforward networks can approximate any static, non-linear, MIMO function, recurrent networks are capable of accurately representing any non-linear dynamic function.

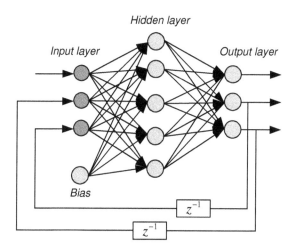

Figure A.2 Jordan recurrent neural network.

The neurons are laid out in layers. The first layer is called the input layer and acts as the sensory organ for the ANN. The various inputs of the ANN are received through this layer. The last layer of the ANN is called the output layer and supplies the ANN output. All the intermediate layers are hidden layers, because their inputs and outputs are not accessible to the external world. The inputs to the hidden layer or output layer neurons are the outputs of previous layer neurons multiplied by the synaptic weights. The structure of a feedforward network is generally denoted by a set of numbers representing the number of inputs, neurons or outputs in each layer. For example, the structure of a four layered network with 10 inputs, 20 neurons in the first hidden layer, 15 neurons in the second hidden layer, and 2 outputs would be denoted by 10-20-15-2.

The process of training an ANN is defined as learning and several algorithms for ANN learning are available in the literature [80]. However the most popular algorithm is known as the backpropagation algorithm. As the name implies, this algorithm modifies the weights of the networks by propagating the errors at the output backwards through the network. The training data is presented to the ANN one data vector at a time, and this is referred to as iteration. The presentation of the whole data set to the ANN is referred to as epoch. The weights are modified at each iteration as in equation (A.1).

$$\Delta w_{ji}(n) = \alpha \Delta w_{ji}(n-1) + \eta \delta_j(n) y_i(n) \qquad (A.1)$$

Where $w_{ji}(n)$ represents the synaptic weight from neuron j in one layer to neuron i in the previous layer at the nth iteration, $y_i(n)$ is the output of neuron i at the nth iteration. α is called the momentum constant and η is called the learning rate. This equation is known as generalized delta rule. There are no fixed rules for choosing the training parameters $(\alpha \ and \ \eta)$. A high value of η makes the ANN converge faster but might lead to instabilities, or might miss the optimum set of weights required for effective learning. A low value of η makes learning slower, but is more stable. However, the training algorithm might get trapped in local minima of the multidimensional weight surface and still not reach an optimum set of weights. It is always good to try out different learning rates for training. The momentum constant α, is less than unity and is usually not too small. It has a smaller effect on network training.

The neuron in any layer computes the weighted sum of the inputs and passes this sum through a non-linear function called the activation function. Usually the sigmoid function is used as the activation function, because it has many desirable properties – it is non-linear and differentiable and its output is limited in an asymptotic fashion.

This function is given by

$$f(x) = \frac{\left[1 - e^{-\beta x}\right]}{\left[1 + e^{\beta x}\right]} \tag{A.2}$$

Where β represents the slope of the activation function. There is another term associated with this sum, and this is called the bias term. In addition to the three inputs, a fixed bias input is also connected through weights to all the hidden and output neurons (all connections are not shown in the figure), for each of the layers other than the input layer, as shown in Figure A.1.

In summary, an ANN is a parallel distributed information processing structure with the following characteristics [81]:

- It is a neurally inspired mathematical model.
- It consists of a large number of highly interconnected processing elements.
- Its connections (weights) hold the knowledge.
- A processing element can dynamically respond to its input stimulus, and the response completely depends on its local information; that is, the input signals arrive at the processing elements via impinging connections and connection weights.
- It has the ability to learn, recall and generalize from training data by assigning or adjusting the connection weights.
- Its collective behaviour demonstrates the computational power, and no single neuron carries specific information.

ANNs have been used successfully in various areas including image processing and recognition, control systems, speech processing, optimization, communication, signal classification, robotics, power systems and many others [81]. ANNs have the ability to approximate almost any continuous non-linear function, and this feature is extremely useful in applications where the functional relationship between inputs and outputs is very complex or unknown. Image and speech processing and financial prediction with ANNs have received widespread attention and commercial products and commercial products are now available which use this technology. Only recently, ANNs have been applied to problems in control systems and this field has yet to reach its maturity.

A.2 Introduction to Fuzzy Logic

Humans use linguistic terms like very hot or slightly old in normal conversations. Although the implied meaning of such linguistic terms is known, it is possible to express this meaning mathematically using conventional set theory. Zadeh introduced Fuzzy sets in his landmark paper in 1965 [82]. The main difference between fuzzy and conventional sets is in the definition of membership of a given set. For conventional set theory, a number either is or is not a member of any given set. The membership can then be represented as a binary quantity: either 1 or 0. On the other hand, the membership of a number in a fuzzy set is defined in terms of a membership function, which generally varies between 0 and 1. The membership function represents a degree of membership of the value in the particular fuzzy set. For example, consider a set of old people. In conventional set theory, one can define such a set as having all people above a certain age, say 65, as members. If person A is 64 years old, he does not belong to this set of old people, while a person B who is 66 years old belongs to this set. Actual linguistic definitions are not so rigid. A fuzzy set of old people can then be defined as a membership function which is zero for any age below 60 and then increases steadily with a value of 1 for any age above 70. Such a membership function is shown in Figure A.3. For this fuzzy set, the membership of person A is 0.4 and that of B is 0.6. Thus, the membership represents a degree to which a person is considered old. Any person with an age above 70 has a membership of 1, indicating that the person is certainly old. Fuzzy set theory lays down a systematic way of dealing with linguistic constructions and defines all operations similar to conventional set theory [83].

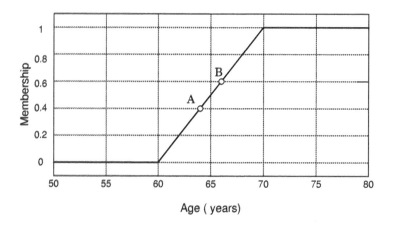

Figure A.3 Example of a membership function for fuzzy sets.

Fuzzy control uses fuzzy set theory to define a non-linear controller and was first developed by Mamdani in 1975 [84]. A fuzzy controller has three components: a fuzzifier, a rule base and a defuzzifier as shown in Figure A.4 [85]. The inputs to the controller are crisp numbers and the outputs are also crisp numbers, but all processing inside the controller is done using fuzzy variables. The first step is to convert the crisp inputs to memberships of each fuzzy set defined for the inputs. This operation is called fuzzification and the block that performs this operation is called a fuzzifier. These membership functions are used in the rule base which relates fuzzy values of the inputs to the output of the rule base. The output of the rule base is also a set of membership values of the fuzzy sets defined for the output variable. To interface with the physical world, this fuzzy variable has to be converted back to a crisp number. This is done by the defuzzifier.

Figure A.4 Components of a fuzzy controller.

A.2.1 Fuzzifier

The fuzzifier converts a crisp input to a membership function of the fuzzy sets defined for that input. Figure A.5 shows an example of membership functions defined for a fuzzy variable. Typically fuzzy sets are named linguistically as Large Negative, Zero or some variation of these names. Given a crisp input value, the membership for each fuzzy set can be found from the membership functions. The figure shows five fuzzy sets labeled *LN, SN, Z, SP*, and *LP* using a triangular membership function. The membership functions are defined over an input range of [-6, 6]. This is generally a normalized range selected for ease of programming. The actual range in terms of physical quantities is dependent on the normalization parameter (scaling factor or gain) for the particular variable. For example, if the error gain is 0.1, an error in the range of [-60, 60] is mapped to the normalized values of [-6, 6]. The membership functions at the two limits (*LN* and *LP*) extend to the physical limits of the variables. Thus, if error is > 60, with an error gain of 0.1, the membership for LP will be 1, with all other sets having zero membership values.

Figure A.5 shows an example of fuzzification. A crisp input of -4 is shown for which two sets have non-zero membership: *LN* and *SN*. The membership of *LN* is about 0.3

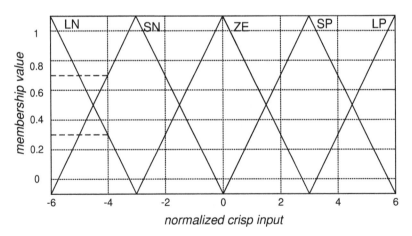

Figure A.5 Sample membership functions used by a fuzzifier.

while that of *SN* is about 0.7. Thus, given the crisp input -4, the fuzzy variable is given by the set of memberships: [0.3 0.7 0 0 0].

A.2.2 Rule base

The rule base maps the fuzzy sets for the inputs to the fuzzy sets for the output and constitutes the core of a fuzzy controller. Table 10 shows a sample rule base. *SN, LN, LP* are notations for fuzzy sets named Small Negative, Large Negative, Large Positive respectively. The rule base has two dimensions corresponding to a fuzzy controller with two inputs and one output. Each row gives the resultant output fuzzy set for each combination of input fuzzy sets. Thus if input 1 is *LN* and input 2 is SP, then the output is *SN* and so on. The entire rule base can be described in terms of such IF-THEN statements or rules, which are OR-ed together. Given the input fuzzy set membership values, the rules are evaluated using the compositional rule of inference to obtain the output fuzzy sets.

Table A.1 Sample rule base for a fuzzy controller

		Input 2				
		LN	LN	LN	LN	LN
Input 1	LN	LN	LN	LN	SN	ZE
	SN	LN	SN	SN	ZE	SP
	ZE	LN	SN	ZE	SP	LP
	SP	SN	ZE	SP	SP	LP
	LP	ZE	SP	LP	LP	LP

Different compositional rules have been defined which use different functions for the AND and OR operations. Let *A* denote the fuzzy variable input 1, *B* denote input 2, *O* denote the output, *a* denote the value of input 1, *b* denote the value of input 2 and *o* denote the value of the output. Then using the minimum function for the AND operation, the rule:

IF *A* is LN AND *B* is SP, THEN *O* is SN, is evaluated by,

$$\mu_o^{SN-k}(o) = \min\left(\mu_A^{LN}(a), \mu_B^{LP}(b)\right) \tag{A.3}$$

Where μ denotes the membership function. The membership function of the output for the set SN is obtained as a result of evaluation of the above rule and is denoted by μ_o^{SN-k}. The superscript *SN_k* denotes that the membership is for the fuzzy set SN as a result of evaluating the kth rule. All rules having the output SN is then combined using the OR operation to get the final membership of the output in the fuzzy set SN. If the maximum function is used for the OR operation, we get:

$$\mu_o^{SN}(o) = \max(\mu_o^{SN-i}) \tag{A.4}$$

The same procedure is adopted for all rules to get all the fuzzy memberships of the output. This min-max inference is the most common although other kinds such as the min-sum inference and the product-sum inference are also widely used.

From the sample rule base, it can be seen that many rules can give the same resulting fuzzy set for the output. The output is SN for four combinations of input 1 and input 2. However, it can be seen from the membership functions in Figure A.5 that for a given crisp value, only two membership functions overlap, so that at most two sets will have non-zero memberships for each input. This implies that at most four rules will be activated simultaneously. This is typical of most fuzzy controllers. The advantages of this characteristic are:

- The rule base can be tuned very easily, since for some given inputs, depending on the performance of the controller, only the active rules need to be adjusted.

This is in contrast to linear controllers, where changing any gain affects the system performance over its entire range of operation.

- The output of the rule base is converted to a crisp value using the defuzzifier. This provides for inherent smoothening of the controller output. Thus, a fuzzy controller is a true non-linear controller.

A.2.3 Defuzzifier

The output of the rule base is a fuzzy variable which is converted to a crisp value by the defuzzifier. Many defuzzification methods have been proposed, of which the centre-of-area or centroid defuzzification is most common. Using this method, the crisp output is given by:

$$o = \frac{\sum_{i=1}^{p} \mu_i A_i c_i}{\sum_{i=1}^{p} \mu_i A_i} \tag{A.5}$$

where, μ_i = membership value of control input for the ith fuzzy set of the output
A_i = area of the membership function for the ith fuzzy set of the output

c_i = centroid of the membership function for the ith fuzzy set of the output
p = number of fuzzy sets for the output

If the areas of all membership functions for the output are the same, the A_i term drops out of the defuzzification equation.

APPENDIX B

INDUCTION MOTOR MODEL AND PARAMETERS

B.1 Dynamic model of the induction motor in stationary reference frame

A three phase induction machine can be represented by an equivalent two phase machine as shown in Figure B.1, where $d^s - q^s$ axes correspond to stator direct and quadrature axes. The dynamic model of the induction motor in the stationary reference frame, used for the analysis carried out in this book has been explained here with equivalent circuits and a series of equations (B.1) to (B.10).

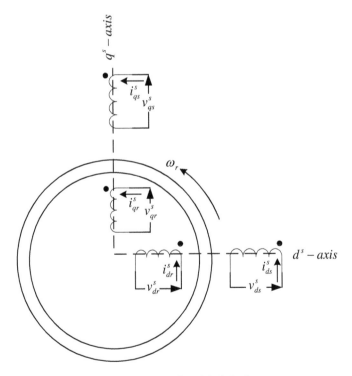

Figure B.1 *d-q* representation of the induction motor.

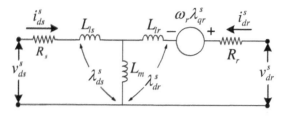

(a) d-axis circuit

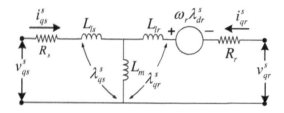

(b) q-axis circuit.

Figure B.2 *dq* equivalent circuit of induction motor in stationary reference frame.

$$v_{ds}^s = R_s i_{ds}^s + \frac{d}{dt}\left(\lambda_{ds}^s\right) \tag{B.1}$$

$$v_{qs}^s = R_s i_{qs}^s + \frac{d}{dt}\left(\lambda_{qs}^s\right) \tag{B.2}$$

$$v_{dr}^s = R_r i_{dr}^s + \frac{d}{dt}\left(\lambda_{dr}^s\right) + \omega_r \lambda_{dr}^s = 0 \tag{B.3}$$

$$v_{qr}^s = R_r i_{qr}^s + \frac{d}{dt}\left(\lambda_{qr}^s\right) - \omega_r \lambda_{qr}^s = 0 \tag{B.4}$$

$$\lambda_{ds}^s = L_{ls} i_{ds}^s + L_m \left(i_{ds}^s + i_{dr}^s\right) \tag{B.5}$$

$$\lambda_{qs}^s = L_{ls} i_{qs}^s + L_m \left(i_{qs}^s + i_{qr}^s\right) \tag{B.6}$$

$$\lambda_{dr}^s = L_{lr} i_{dr}^s + L_m \left(i_{ds}^s + i_{dr}^s\right) \tag{B.7}$$

$$\lambda_{qr}^s = L_{lr} i_{qr}^s + L_m \left(i_{qs}^s + i_{qr}^s\right) \tag{B.8}$$

$$T_e = \frac{3}{2} p L_m \left(i_{qs}^s i_{dr}^s - i_{ds}^s i_{qr}^s\right) \tag{B.9}$$

$$J_T \frac{d\omega_m}{dt} = \frac{1}{p} J_T \frac{d\omega_r}{dt} = T_e - T_L \tag{B.10}$$

B.2 Dynamic model of the induction motor in synchronously rotating reference frame

The induction motor equations described in B.1 are referred to the stationary reference frame. These equations can also be referred to the synchronously rotating reference aligned with *d-q* axes shown in Figure 2.1.

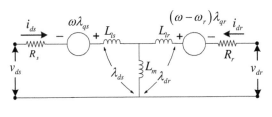

(a) *d*-axis circuit

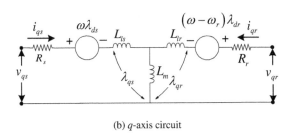

(b) *q*-axis circuit

Figure B.3 *d-q* equivalent circuit of induction motor in synchronously rotating reference frame.

When equations (B.1) and (B.2) are referred to the synchronously rotating reference frame rotating at ω,

$$v_{ds} = R_s i_{ds} + \frac{d}{dt}(\lambda_{ds}) - \omega\lambda_{qs} \tag{B.11}$$

$$v_{qs} = R_s i_{qs} + \frac{d}{dt}(\lambda_{qs}) + \omega\lambda_{ds} \tag{B.12}$$

The last terms in equations are defined as speed emfs due to the rotation of axes. That is, when $\omega = 0$, the equations revert to the stationary reference frame form.

The equations (B.3) and (B.4) for the rotor become:

$$v_{dr} = R_r i_{dr} + \frac{d}{dt}(\lambda_{dr}) - (\omega - \omega_r)\lambda_{dr} = 0 \tag{B.13}$$

$$v_{qr} = R_r i_{qr} + \frac{d}{dt}(\lambda_{qr}) + (\omega - \omega_r)\lambda_{dr} = 0 \tag{B.14}$$

All the sinusoidal variables in the stationary frame appear as dc quantities in the synchronously rotating reference frame.

The flux linkage expressions can be written as:

$$\lambda_{ds} = L_{ls} i_{ds} + L_m \left(i_{ds} + i_{dr} \right) \tag{B.15}$$

$$\lambda_{qs} = L_{ls} i_{qs} + L_m \left(i_{qs} + i_{qr} \right) \tag{B.16}$$

$$\lambda_{dr} = L_{ls} i_{dr} + L_m \left(i_{ds} + i_{dr} \right) \tag{B.17}$$

$$\lambda_{qr} = L_{ls} i_{qr} + L_m \left(i_{qs} + i_{qr} \right) \tag{B.18}$$

The torque expression of the motor can be derived as follows:

$$T_e = \frac{3}{2} p L_m \left(i_{qs} i_{dr} - i_{ds} i_{qr} \right) \tag{B.19}$$

$$= \frac{3}{2} p \left(\lambda_{dr} i_{qr} - \lambda_{qr} i_{dr} \right) \tag{B.20}$$

B.3   Determination of electrical parameters

Open circuit and short circuit tests were conducted to find out the parameters of the induction motor. The stator resistance R_s was obtained with a DC measurement.

The open-circuit test was conducted by supplying the rated voltage to the stator winding while driving the induction motor at its synchronous speed using the DC motor coupled to the induction motor. When the motor runs at synchronous speed, the slip will be zero and hence the rotor current becomes zero. The conventional equivalent circuit model for this test can be reduced to the one shown in Figure B.4.

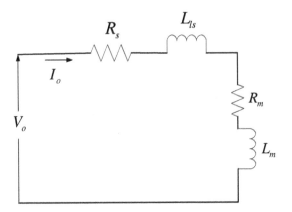

Figure B.4 Per-phase equivalent circuit under no-load test.

Where, V_o – the measured open-circuit phase voltage

I_o – the measured open-circuit phase current

P_o – the measured open-circuit three phase power

Total input resistance $R_o = \dfrac{P_o}{3I_o^{\,2}}$ (B.21)

Total input impedance $Z_o = \dfrac{V_o}{I_o}$ (B.22)

Total input reactance $X_o = \sqrt{Z_o^{\,2} - R_o^{\,2}}$ (B.23)

Then, $R_m = R_o - R_s$ (B.24)

$X_m = X_o - X_{ls}$ (B.25)

The locked rotor test was conducted by blocking the rotor and supplying the rated current from a 50 Hz ac supply.

Under blocked rotor condition, as the slip is equal to 1.0, the conventional equivalent circuit with the magnetizing branch neglected, will be reduced to the one shown in Figure B.5.

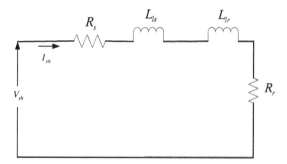

Figure B.5 Per-phase equivalent circuit under blocked rotor test.

Total input resistance $R_{sh} = \dfrac{P_{sh}}{3I_{sh}^2}$ (B.26)

Total input impedance $Z_{sh} = \dfrac{V_{sh}}{I_{sh}} = R_{sh} + jX_{sh}$ (B.27)

Total input reactance $X_{sh} = \sqrt{Z_{sh}^2 - R_{sh}^2}$ (B.28)

Where, V_{sh} – the measured short-circuit phase voltage

I_{sh} – the measured short-circuit phase current

P_{sh} – the measured short-circuit three phase input power

Now the rotor resistance can be calculated as $R_r = R_{sh} - R_s$ (B.29)

The total leakage inductance $(L_{ls} + L_{lr})$ has been split into two equal halves as indicated in Table I. Various other ratios have also been investigated. It was found that other ratios (up to a factor of four) do not appreciably influence the results which are obtained by splitting of the leakage inductance equally. The total leakage inductance being smaller by a factor of nearly nine, than the mutual inductance L_m, this result is expected.

$$X_{ls} = X_{lr} = X_{sh}/2 \qquad\qquad (B.30)$$

B.4 Determination of mechanical parameters J and D of the induction motor set-up

The J and D of the induction motor can be determined by using the coupled permanent magnet DC motor. The following equations can be derived for the DC machine:

$$E_a = V - I_a R_a \tag{B.31}$$

$$P = E_a I_a \tag{B.32}$$

$$E_a = k\omega \tag{B.33}$$

$$T_e = k I_a \tag{B.34}$$

$$T_e = J\frac{d\omega}{dt} + D\omega + T_L \tag{B.35}$$

$$T_e - T_L = D\omega \tag{B.36}$$

In steady-state, if the load torque shown in (B.26) is equal to zero, (B.27) can be written as:

$$\frac{E_a I_a}{\omega} = D\omega \tag{B.37}$$

$$P = D\omega^2 \tag{B.38}$$

When the DC motor was driving the induction motor as load, the voltage, current and rotor speed were recorded and the power was calculated as in (B.32). The friction coefficient D was calculated from (B.38).

Where E_a – back-emf

         V – DC voltage

         I_a – DC motor current

         ω - speed in rad/second

         T_e – electromagnetic torque

         T_L – load torque

         D – friction coefficient

         J – rotor inertia

A. 1.1 kW Squirrel Cage induction motor

In order to calculate the friction coefficient, the speed of the permanent magnet DC motor was varied by controlling the DC input voltage. The input power was plotted against ω^2, where ω is the speed in radians/second. This plot is shown in Figures B.5. From the slope of the plot in Figure B.5, the friction coefficient D was calculated to be 0.0027 Nm/rad/second.

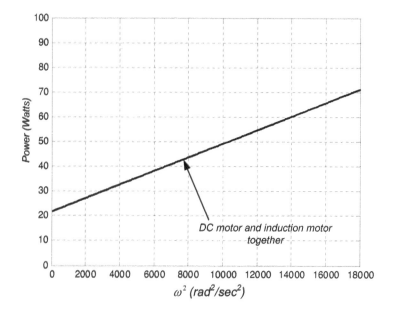

Figure B.6 Power vs. speed plot for the squirrel-cage induction motor together with DC motor.

To calculate the total inertia J of the induction motor- DC motor set-up, a run-down test was conducted of the DC motor. Initially, the input voltage to the DC motor was adjusted so that it runs at nearly 1200 rpm, and the input power measured. The input voltage was switched off and the speed was recorded until the motor comes to rest. When the electro-magnetic torque generated by the DC motor is equal to zero, equation (B.35) can be written as:

$$0 = J\frac{d\omega}{dt} + D\omega \tag{B.39}$$

From (B.39), J can be calculated as:

$$J = -D\omega\big|_{t=0}\Big/\frac{d\omega}{dt}\bigg|_{t=0} \tag{B.40}$$

The plot of speed vs. time for the induction motor together with the permanent magnet DC motor obtained from a run-down test has been shown in Figure B.6. From this plot, the combined rotor inertia of the induction motor together with the DC motor has been calculated to be 0.0117kg.m^2 using Equation (B.40).

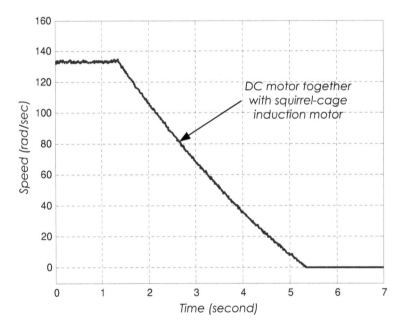

Figure B.7 Run-down speed-time plot for the squirrel-cage induction motor together with permanent magnet DC motor.

All the parameters of the induction motor used for control and estimation are shown in Table B.1

TABLE B.I.   SQUIRREL-CAGE INDUCTION MOTOR PARAMETERS

Type	3 Phase Y-connected squirrel cage (Made by ABB)
Rated Power	1.1 kW
Rated Voltage	415 V
Rated current	2.77 Amps
Rated frequency	50 Hz
Rated speed	1415 rpm
Number of poles	4
Stator Resistance R_s	6.03 Ω
Rotor Leakage Inductance L_{lr}	29.9 mH
Stator Leakage Inductance L_{ls}	29.9 mH
Magnetizing Inductance L_m	489.3 mH
Rotor Resistance R_r	6.085 Ω at 50 Hz
Moment of Inertia J_T	0.011787 kgm^2
Friction Coefficient D	0.0027 Nm/rad/second

B. 3.6 kW Slip-ring induction motor

In order to calculate the friction coefficient of the slip-ring motor – DC motor set-up, the speed of the DC motor was varied by controlling the DC input voltage. The input power was plotted against ω^2, where ω is the speed in radians/second. This plot is shown in Figures B.8. From the slope of the plot in Figure B.8, the friction coefficient D has been calculated to be 0.0089 Nm/rad/second.

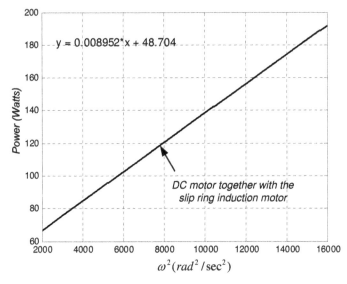

Figure B.8 Power vs. speed plot for the slip-ring induction motor together with DC motor.

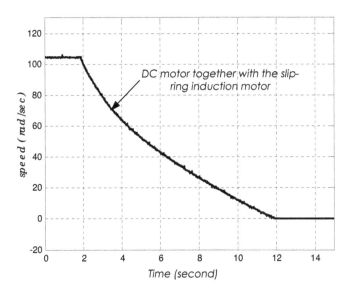

Figure B.9 Run-down speed-time plot for the slip-ring induction motor together with DC motor.

To determine the total inertia J of the slip ring induction motor – DC motor set-up, a run-down test was conducted as explained for the squirrel cage induction motor. The plot of speed vs. time for the slip-ring induction motor together with the DC motor obtained from a run-down test is shown in Figure B.9. From this plot, the combined rotor inertia of the induction motor together with the DC motor has been calculated to be 0.0421kg.m^2 using Equation (B.40).

TABLE B.II.   SLIP-RING INDUCTION MOTOR PARAMETERS

Type	3 Phase Y-connected slip ring (Made by AEG)
Rated Power	3.6 kW
Rated Voltage	415V
Rated current	7.8 Amps
Rated frequency	50 Hz
Rated speed	1410 rpm
Number of poles	4
Stator Resistance R_s	1.54 Ω
Rotor Leakage Inductance L_{lr}	11.65 mH
Stator Leakage Inductance L_{ls}	11.65 mH
Magnetizing Inductance L_m	184.61 mH
Rotor Resistance R_r	2.7 Ω at 50 Hz
Moment of Inertia J_T	0.0421 kgm^2
Friction Coefficient D	0.0089 Nm/rad/second

APPENDIX C

ANALYSIS OF STATOR CURRENT ESTIMATION

The discrete equations used for estimating the stator current from voltage and current models of the induction motor is derived here. These equations have been referred to in Chapters 6, 7 and 8.

The voltage model of the induction motor can be written as:

$$
\begin{bmatrix} \dfrac{d\lambda_{dr}^{s\,vm}}{dt} \\[2mm] \dfrac{d\lambda_{qr}^{s\,vm}}{dt} \end{bmatrix} = \frac{L_r}{L_m} \left\{ \begin{bmatrix} v_{ds}^{s} \\ v_{qs}^{s} \end{bmatrix} - R_s \begin{bmatrix} i_{ds}^{s} \\ i_{qs}^{s} \end{bmatrix} - \sigma L_s \begin{bmatrix} \dfrac{di_{ds}^{s}}{dt} \\[2mm] \dfrac{di_{qs}^{s}}{dt} \end{bmatrix} \right\}
\tag{C.1}
$$

The current model of the induction motor can be written as:

$$
\begin{bmatrix} \dfrac{d\lambda_{dr}^{s\,im}}{dt} \\[2mm] \dfrac{d\lambda_{qr}^{s\,im}}{dt} \end{bmatrix} = \begin{bmatrix} -\dfrac{1}{T_r} & -\omega_r \\[2mm] \omega_r & -\dfrac{1}{T_r} \end{bmatrix} \begin{bmatrix} \lambda_{dr}^{s\,im} \\ \lambda_{qr}^{s\,im} \end{bmatrix} + \frac{L_m}{T_r} \begin{bmatrix} i_{ds}^{s} \\ i_{qs}^{s} \end{bmatrix}
\tag{C.2}
$$

From (C.1),

$$
\frac{d\lambda_{dr}^{s\,vm}}{dt} = \frac{L_r}{L_m} \left[v_{ds}^{s} - R_s i_{ds}^{s} - \sigma L_s \frac{di_{ds}^{s}}{dt} \right]
\tag{C.3}
$$

Re-arranging the terms in (C.3),

$$
\frac{L_m}{L_r} \frac{d\lambda_{dr}^{s\,vm}}{dt} = v_{ds}^{s} - R_s i_{ds}^{s} - \sigma L_s \frac{di_{ds}^{s}}{dt}
\tag{C.4}
$$

$$
\sigma L_s \frac{di_{ds}^{s}}{dt} = v_{ds}^{s} - R_s i_{ds}^{s} - \frac{L_m}{L_r} \frac{d\lambda_{dr}^{s\,vm}}{dt}
\tag{C.5}
$$

The current model equation can be written as in (C.3).

From the current model equations (C.2),

$$\frac{d\lambda_{dr}^{s\,im}}{dt} = -\frac{1}{T_r}\lambda_{dr}^{s\,im} - \omega_r\lambda_{qr}^{s\,im} + \frac{L_m}{T_r}i_{ds}^s \qquad (C.6)$$

Substituting (C.6) in (C.5),

$$\sigma L_s\frac{di_{ds}^s}{dt} = v_{ds}^s - R_s i_{ds}^s - \frac{L_m}{L_r}\left[-\frac{1}{T_r}\lambda_{dr}^{s\,im} - \omega_r\lambda_{qr}^{s\,im} + \frac{L_m}{T_r}i_{ds}^s\right] \qquad (C.7)$$

$$\sigma L_s\frac{di_{ds}^s}{dt} = v_{ds}^s - R_s i_{ds}^s + \frac{L_m}{L_r^2}\frac{1}{R_r}\lambda_{dr}^{s\,im} + \frac{L_m}{L_r}\omega_r\lambda_{qr}^{s\,im} - \frac{L_m^2}{L_r^2}\frac{1}{R_r}i_{ds}^s \qquad (C.8)$$

$$\sigma L_s\frac{di_{ds}^s}{dt} = \frac{L_m}{L_r^2}\frac{1}{R_r}\lambda_{dr}^{s\,im} + \frac{L_m}{L_r}\omega_r\lambda_{qr}^{s\,im} - \frac{L_m^2}{L_r^2}\frac{1}{R_r}i_{ds}^s + v_{ds}^s - R_s i_{ds}^s \qquad (C.9)$$

Similarly from (C.1) it can also be written:

$$\frac{d\lambda_{qr}^{s\,vm}}{dt} = \frac{L_r}{L_m}\left[v_{qs}^s - R_s i_{qs}^s - \sigma L_s\frac{di_{qs}^s}{dt}\right] \qquad (C.10)$$

Re-arranging the terms in (C.10):

$$\frac{L_m}{L_r}\frac{d\lambda_{qr}^{s\,vm}}{dt} = v_{qs}^s - R_s i_{qs}^s - \sigma L_s\frac{di_{qs}^s}{dt} \qquad (C.11)$$

$$\sigma L_s\frac{di_{qs}^s}{dt} = v_{qs}^s - R_s i_{qs}^s - \frac{L_m}{L_r}\frac{d\lambda_{qr}^{s\,vm}}{dt} \qquad (C.12)$$

From the current model equations (C.2),

$$\frac{d\lambda_{qr}^{s\,im}}{dt} = \omega_r\lambda_{dr}^{s\,im} - \frac{1}{T_r}\lambda_{qr}^{s\,im} + \frac{L_m}{T_r}i_{qs}^s \qquad (C.13)$$

Substituting (C.13) in (C.12),

$$\sigma L_s\frac{di_{qs}^s}{dt} = v_{qs}^s - R_s i_{qs}^s - \frac{L_m}{L_r}\left[\omega_r\lambda_{dr}^{s\,im} - \frac{1}{T_r}\lambda_{qr}^{s\,im} + \frac{L_m}{T_r}i_{qs}^s\right] \qquad (C.14)$$

$$\sigma L_s\frac{di_{qs}^s}{dt} = v_{qs}^s - R_s i_{qs}^s + \frac{L_m}{L_r^2}\frac{1}{R_r}\lambda_{qr}^{s\,im} - \frac{L_m}{L_r}\omega_r\lambda_{dr}^{s\,im} - \frac{L_m^2}{L_r^2}\frac{1}{R_r}i_{qs}^s \qquad (C.15)$$

$$\sigma L_s\frac{di_{qs}^s}{dt} = \frac{L_m}{L_r^2}\frac{1}{R_r}\lambda_{qr}^{s\,im} - \frac{L_m}{L_r}\omega_r\lambda_{dr}^{s\,im} - \frac{L_m^2}{L_r^2}\frac{1}{R_r}i_{qs}^s + v_{qs}^s - R_s i_{qs}^s \qquad (C.16)$$

Using the discrete model of equation (C.9) for a sampling time T_S :

$$\sigma L_s \left[\frac{i_{ds}^{s\,*}(k) - i_{ds}^{s\,*}(k-1)}{T_s} \right] = \frac{L_m}{L_r^2} \frac{1}{R_r} \lambda_{dr}^{s\,im}(k-1) + \frac{L_m}{L_r} \omega_r \lambda_{qr}^{s\,im}(k-1)$$

$$- \frac{L_m^2}{L_r^2} \frac{1}{R_r} i_{ds}^{s\,*}(k-1) + v_{ds}^s(k-1) - R_s i_{ds}^s(k-1) \quad (C.17)$$

Re-arranging the terms in (C.17),

$$i_{ds}^{s\,*}(k) - i_{ds}^{s\,*}(k-1) = \frac{T_s}{\sigma L_s} \frac{L_m}{L_r^2} \frac{1}{R_r} \lambda_{dr}^{s\,im}(k-1) + \frac{T_s}{\sigma L_s} \frac{L_m}{L_r} \omega_r \lambda_{qr}^{s\,im}(k-1)$$

$$- \frac{T_s}{\sigma L_s} \frac{L_m^2}{L_r^2} \frac{1}{R_r} i_{ds}^{s\,*}(k-1) + \frac{T_s}{\sigma L_s} v_{ds}^s(k-1) - \frac{T_s}{\sigma L_s} R_s i_{ds}^{s\,*}(k-1) \quad (C.18)$$

$$i_{ds}^{s\,*}(k) = i_{ds}^{s\,*}(k-1) + \frac{T_s}{\sigma L_s} \frac{L_m}{L_r^2} \frac{1}{R_r} \lambda_{dr}^{s\,im}(k-1) + \frac{T_s}{\sigma L_s} \frac{L_m}{L_r} \omega_r \lambda_{qr}^{s\,im}(k-1)$$

$$- \frac{T_s}{\sigma L_s} \frac{L_m^2}{L_r^2} \frac{1}{R_r} i_{ds}^{s\,*}(k-1) + \frac{T_s}{\sigma L_s} v_{ds}^s(k-1) - \frac{T_s}{\sigma L_s} R_s i_{ds}^{s\,*}(k-1) \quad (C.19)$$

$$i_{ds}^{s\,*}(k) = \left[1 - \frac{T_s}{\sigma L_s} \frac{L_m^2}{L_r^2} \frac{1}{R_r} - \frac{T_s}{\sigma L_s} R_s \right] i_{ds}^{s\,*}(k-1) + \frac{T_s}{\sigma L_s} \frac{L_m}{L_r^2} \frac{1}{R_r} \lambda_{dr}^{s\,im}(k-1)$$

$$+ \frac{T_s}{\sigma L_s} \frac{L_m}{L_r} \omega_r \lambda_{qr}^{s\,im}(k-1) + \frac{T_s}{\sigma L_s} v_{ds}^s(k-1) \quad (C.20)$$

$$i_{ds}^{s\,*}(k) = W_4 i_{ds}^{s\,*}(k-1) + W_5 \lambda_{dr}^{s\,im}(k-1) + W_6 \omega_r \lambda_{qr}^{s\,im}(k-1) + W_7 v_{ds}^s(k-1) \quad (C.21)$$

where, $\quad W_4 = 1 - \dfrac{T_s}{\sigma L_s} \dfrac{L_m^2}{L_r^2} \dfrac{1}{R_r} - \dfrac{T_s}{\sigma L_s} R_s$

$$W_5 = \frac{T_s}{\sigma L_s} \frac{L_m}{L_r^2} \frac{1}{R_r}$$

$$W_6 = \frac{T_s}{\sigma L_s} \frac{L_m}{L_r}$$

$$W_7 = \frac{T_s}{\sigma L_s}$$

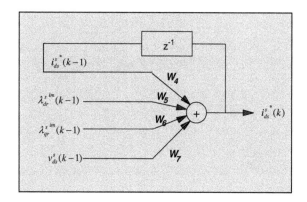

Figure C.1 *d*-axis stator current estimation using recurrent neural network based on (C.21).

Equation (C.16) can also be written as:

$$\sigma L_s \left[\frac{i_{qs}^{s\,*}(k) - i_{qs}^{s\,*}(k-1)}{T_s} \right] = \frac{L_m}{L_r^2} \frac{1}{R_r} \lambda_{qr}^{s\,im}(k-1) - \frac{L_m}{L_r} \omega_r \lambda_{dr}^{s\,im}(k-1)$$

$$- \frac{L_m^2}{L_r^2} \frac{1}{R_r} i_{qs}^{s\,*}(k-1) + v_{qs}^s(k-1) - R_s i_{qs}^{s\,*}(k-1) \quad \text{(C.22)}$$

$$i_{qs}^{s\,*}(k) - i_{qs}^{s\,*}(k-1) = \frac{T_s}{\sigma L_s} \frac{L_m}{L_r^2} \frac{1}{R_r} \lambda_{qr}^{s\,im}(k-1) - \frac{T_s}{\sigma L_s} \frac{L_m}{L_r} \omega_r \lambda_{dr}^{s\,im}(k-1)$$

$$- \frac{T_s}{\sigma L_s} \frac{L_m^2}{L_r^2} \frac{1}{R_r} i_{qs}^{s\,*}(k-1) + \frac{T_s}{\sigma L_s} v_{qs}^s(k-1)$$

$$- \frac{T_s}{\sigma L_s} R_s i_{qs}^{s\,*}(k-1) \quad \text{(C.23)}$$

$$i_{qs}^{s\,*}(k) = i_{qs}^{s\,*}(k-1) + \frac{T_s}{\sigma L_s} \frac{L_m}{L_r^2} \frac{1}{R_r} \lambda_{qr}^{s\,im}(k-1) - \frac{T_s}{\sigma L_s} \frac{L_m}{L_r} \omega_r \lambda_{dr}^{s\,im}(k-1)$$

$$- \frac{T_s}{\sigma L_s} \frac{L_m^2}{L_r^2} \frac{1}{R_r} i_{qs}^{s\,*}(k-1) + \frac{T_s}{\sigma L_s} v_{qs}^s(k-1)$$

$$- \frac{T_s}{\sigma L_s} R_s i_{qs}^{s\,*}(k-1) \quad \text{(C.24)}$$

$$i_{qs}^{s\,*}(k) = \left[1 - \frac{T_s}{\sigma L_s}\frac{L_m^2}{L_r^2}\frac{1}{R_r} - \frac{T_s}{\sigma L_s}R_s\right]i_{qs}^{s\,*}(k-1) + \frac{T_s}{\sigma L_s}\frac{L_m}{L_r^2}\frac{1}{R_r}\lambda_{qr}^{s\,im}(k-1)$$

$$-\frac{T_s}{\sigma L_s}\frac{L_m}{L_r}\omega_r\lambda_{dr}^s(k-1) + \frac{T_s}{\sigma L_s}v_{qs}^s(k-1) \qquad (C.25)$$

$$i_{qs}^{s\,*}(k) = W_4 i_{qs}^{s\,*}(k-1) + W_5\lambda_{qr}^{s\,im}(k-1) - W_6\omega_r\lambda_{dr}^{s\,im}(k-1) + W_7 v_{qs}^s(k-1) \qquad (C.26)$$

where, $\quad W_4 = 1 - \dfrac{T_s}{\sigma L_s}\dfrac{L_m^2}{L_r^2}\dfrac{1}{R_r} - \dfrac{T_s}{\sigma L_s}R_s$

$$W_5 = \frac{T_s}{\sigma L_s}\frac{L_m}{L_r^2}\frac{1}{R_r}$$

$$W_6 = \frac{T_s}{\sigma L_s}\frac{L_m}{L_r}$$

$$W_7 = \frac{T_s}{\sigma L_s}$$

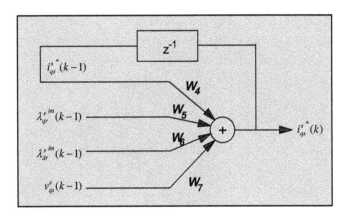

Figure C.2 q-axis stator current estimation using recurrent neural network based on (C.26).

APPENDIX D

EXPERIMENTAL SET-UP

An experimental set-up was built for a squirrel-cage induction motor during the course of this study and is shown in Figure D.1. The motor generator set consists of a 1.1kW three phase induction motor coupled to 1.1kW permanent magnet DC motor. An encoder with 5000 pulses/rev was mounted to the induction motor shaft. A PC with Pentium III 800MHz was used to host the dSPACE controller board and the control software development throughout this project. During the initial stages, while developing the RFOC drive in chapter 4 and the rotor flux estimation using the programmable cascaded filter in chapter 5, the DS1102 control board was used as

Figure D.1 Experimental set-up of the 1.1kW squirrel-cage induction motor drive.

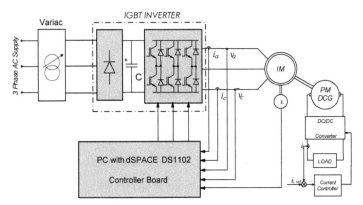

Figure D.2 Experimental set-up of the 1.1kW squirrel-cage induction motor drive with DS1102 Controller.

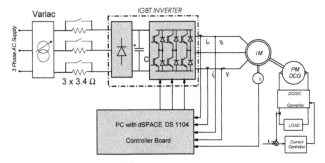

Figure D.3 Experimental set-up of the 1.1kW squirrel-cage induction motor drive with DS1104 Controller.

shown in Figure D.2. The experimental results for the remaining chapters was carried out with a faster DS1104 control board as presented in Figure D.3. A 5kW IGBT inverter using Mitsubishi IPM with necessary voltage and current sensor boards was used to drive the induction motor. A three phase variac was used to adjust the input voltage to the inverter. A permanent magnet DC motor coupled to the induction motor was used to load the induction motor. A constant load torque could be maintained by using the current control loop in the load circuit. The current control in the load motor was achieved using a dedicated hardware current controller board and a DC/DC converter.

Subsequently, it was necessary to verify the rotor resistance estimation with a slip-ring induction motor. To achieve this, another experimental set-up was built for a 3.6kW slip-ring induction motor as shown in Figure D.4. The 5kW IGBT inverter and PC with DS1104 board are used in both set-ups.

Figure D.4 Experimental set-up of the 3.6kW slip-ring induction motor drive.

www.ingramcontent.com/pod-product-compliance
Lightning Source LLC
LaVergne TN
LVHW042333060326
832902LV00006B/141